HOW TO BE A HUMANKIND SUPERHERO

A Manifesto for Individuals to Reclaim a Safe Climate

HOW TO BE A HUMANKIND SUPERHERO

A Manifesto for Individuals to Reclaim a Safe Climate

Harold Forbes

Copyright © 2010 Harold Forbes

The moral right of the author has been asserted.

Apart from any fair dealing for the purposes of research or private study, or criticism or review, as permitted under the Copyright, Designs and Patents Act 1988, this publication may only be reproduced, stored or transmitted, in any form or by any means, with the prior permission in writing of the publishers, or in the case of reprographic reproduction in accordance with the terms of licences issued by the Copyright Licensing Agency. Enquiries concerning reproduction outside those terms should be sent to the publishers.

Every effort has been made to fulfil the requirements with regard to reproducing copyright material. The author and publisher will be glad to rectify and omissions at the earliest opportunity.

Matador
5 Weir Road
Kibworth Beauchamp
Leicester LE8 0LQ UK
Tel: 0116 279 2299
Email: books@troubador.co.uk
Web: www.troubador.co.uk/matador

ISBN 978-1848764-033

Typeset in 11pt Century Gothic by Troubador Publishing Ltd, Leicester, UK
Printed in the UK by TJ International Ltd, Padstow, Cornwall

Matador is an imprint of Troubador Publishing Ltd

This book is dedicated to the memory of my mother, Margaret Russell 'Peggy' Forbes (12th July 1926 to 23rd March 2008), whose life's work made it possible, and my father, Harold Rutherford Connon Forbes (16th November 1926 to 25th December 1971), the man she loved to her dying day.

Contents

Foreword	ix
Part One	**1**
How We Managed to Get into this Situation	1
Part Two	**29**
The Labours	
Deciding to Act	
(The Labour of The Nemean Lion)	29
Transport	
(The Labour of The Stymphalian Birds)	41
Recycling	
(The Labour of The Augeian Stables)	54
Powering Your House	
(The Labour of The Ceryneian Hind)	58
Using Your House as a Generator and Keeping it Warm	
(The Labour of The Apples of the Hesperides)	65
Food	
(The Labour of The Mares of Diomedes)	85
The Shopping Habit	
(The Labour of The Cretan Bull)	96
Plant Trees	
(The Labour of The Erymanthian Boar)	104

Recruit a Friend or Neighbour to the Cause
(The Labour of The Amazon Queen's Girdle) 114
Shaping Your Local Area
(The Labour of The Cattle of Geryon) 128
Encourage a Company
(The Labour of The Lernaean Hydra) 139

Encourage a Government
(The Labour of Capturing Cerberus) 157

Part Three **179**
Where to from Here?

References 191
Bibliography 196
Appendixes 197

Acknowledgements 201

Foreword

This book has been written because I felt it needed to be written. It needed to be written so I could add my voice and encourage others to join me, so we can help speed action on greenhouse gas emissions; a universal threat much more potent and more likely to lead to mutually assured destruction than deaths from thermonuclear devices, AIDS, bird or swine flu, or even asteroid strikes.

It is not a new problem. The greenhouse effect was first described by Joseph Fourier in 1824.

By 1908, Svante Arrhenius had calculated the relationship between the concentration of carbon dioxide in the atmosphere and temperature, and predicted that human activities could bring about a change to the atmosphere sufficient to cause global warming.

Thirty years ago, the US National Academy of Science published 'Carbon Dioxide and Climate: A Scientific Assessment'[1], which warned: 'If carbon dioxide continues to increase, the study group finds no reason to doubt that climate changes will result and no reason to believe that these changes will be negligible.' (Or put another way: 'As carbon dioxide continues to rise, the climate will change and the effects will be big.')

In 1992, the Rio Earth Summit saw the start of 'international action' on climate change when 189 countries signed up to the United Nations Framework Convention on Climate Change (UNFCCC).

In 1997, the countries negotiated a set of legally-binding emissions cuts called the Kyoto Protocol, although it didn't come into force until 2005 and only runs until 2012.

In 2007, in Bali, the countries agreed that extended action would be necessary and that it would be decided upon in Copenhagen in December 2009. In the event, all that could be agreed upon in Copenhagen was that the world will work to limit the increase in global temperatures to $2°$ Celsius, but they set no targets or timescales for how and by whom reductions in emissions will be made. That will be negotiated at a later date. Meanwhile, greenhouse gas emissions have and will continue to rise.

I also want to express what I believe is a simpler solution to the core problem we face: how do we find enough fuel to provide the energy that our entire civilisation has come to depend upon without causing catastrophic damage to the atmosphere that we are also dependent upon.

Currently, the world fuel strategy until 2030 will be reliant upon oil from fields that are expected to be discovered, and thereafter from 'clean coal' using a technology that is expected to be developed. While the oil industry has a long history of drilling 'dry' holes and might reasonably expect to continue their historic hit rate, the coal industry has been experimenting with 'clean coal' for a number of years without success. I may be a lay person, but neither of these options seems sensible because they do not address the core issue: burning fossil fuels produces poisonous gases and continuing to do so is putting us at great danger as a species.

There is a huge amount of information, debate, opinion and argument about what alternatives 'they' should implement, but it is really difficult to work out how you can best influence the outcome if you decide to join the fight against climate change. My intention is to provide a focus for action that will allow you to make an impact where it matters and avoid getting sidetracked by tokenism or unfocused good intentions.

In the past few years there has been a strong movement to ban plastic bags or at least persuade people to give them up voluntarily. It has been very successful and by May 2009, Marks & Spencer reported an 85% reduction in their use of plastic bags and DEFRA, the Department for Environment, Food and Rural Affairs, stated that retailers were on track to bring about a 5 billion reduction in plastic bag usage, which would be 'equivalent to taking 41,000 cars off the road'.[2]

This is a great illustration of how people want to do something about the environment and an example of how change can come about when multiple parties come together around a simple objective. Unfortunately, 41,000 cars represent just 0.1% of the vehicles currently on UK roads, so it has been a huge effort for little result in carbon terms.

By undertaking the actions in this book you will be tackling the bigger areas where noticeable impacts can be made, allowing you to reduce your personal carbon footprint by between 20 and 50%. You will also contribute to building an unstoppable force for change. These include a mix of actions you can take in your personal life and lobbying of government and business. You must do both.

When President Obama arrived in Copenhagen, he said: 'The time has come for us to get off the sidelines and shape the future that we seek.' I do not know what kind of future you seek for yourself or your family, but I do know that without concerted and impactful action from people around the world, we will be trying to shape our futures in an increasingly hostile and possibly deadly environment. That is why we must solve the climate change problem.

PART ONE

How we managed to get into this situation

Are our animal instincts of greed and selfishness capable of being overcome by our human traits of love and co-operation? After all, we are an amazing bunch, we humans. Our genes have enabled us to climb to the top of the evolutionary pile. Our big brains have allowed us to shape and mould our environment. We have developed complex social systems that allow us to live mostly in harmony and our ability to communicate and interact has allowed us to generate and share knowledge not just with each other but across generations.

We are pretty smart, but even the smartest of us cannot assimilate all of the knowledge we have generated. As a species, we carry on with our quest for ever greater understanding but we have no idea of how much smarter we could be.*

•As US Defence Secretary Donald Rumsfeld said in a press conference in 2002 about the Afghanistan conflict: 'There are known knowns. These are things we know that we know. There are known unknowns. That is to say, there are things that we now know we don't know. But there are also unknown unknowns. These are things we do not know we don't know.' The speech was pretty much ridiculed, which was a pity as it was possibly the only true thing he said during his term of office).

The most amazing thing about us though is that we share much of our make-up with other animals and even plants. We humans share 98% of our DNA, the building blocks of life, with chimpanzees, and even 50% with bananas.[3] Clearly, small differences can sometimes have a big impact.

Genetically, each and every one of us is different from everyone else alive and everyone else who has ever been alive.[4] Yet we share a few immutable characteristics:

- We all need clean water and air to survive, but everything else is a choice (and yes, we do need to eat but the actual food that we choose is pretty varied, and there are lots of things that one individual may never choose to eat but others gorge upon).
- We experience the world in our own way and how we react to our experiences is a big driver of our behaviour.
- How we experience others and form ourselves into groups is critical to our survival.
- None of us can predict the future with a guarantee of accuracy, even though many claim to be able to do so.

These characteristics drive the complex interactions we all deal with on a daily basis. Individually unique we may be, but we all share a similar journey. We are born helpless and only partially formed. We must be nourished, usually by our mother, and cared for over an extended period before we can function as independent beings. Acquiring our teeth and the ability to control our bodily functions can be fraught experiences, certainly for the parent and maybe for the child. As our memory doesn't develop until that stage has passed, it is difficult for us to recall what that experience must have been like but the work of Freud, Erikson and others indicates that our future relationships can be greatly affected by the process.

Nature versus nurture remains a contentious issue. We have inbuilt reactions that are remnants from 'flight or fight' instincts that our ancestors needed to survive but, how much an individual's personality and behaviour is driven by their genes and how much is the result of their conscious or unconscious decisions is still a matter of conjecture. However, it does not prevent each of us waging our own internal struggle between our animal instincts of greed and selfishness and our human traits of love and co-operation.

As we grow physically, we also grow mentally. The bounds of what must and what must not happen to us to ensure our physical growth are fairly clearly defined. The bounds of what *may* influence our mental development are much less clear and have provided fertile ground for philosophers. Our desire to understand them has even generated new disciplines of psychology and psychotherapy.

We do, however, develop a sense of 'self' or a mind that is distinct from our bodies. Like our DNA, our own self is unique but greatly similar to that of others, especially those with whom we live. Our families and our societies are important influences on how we learn to experience the world and react to how we think it treats us. Some thinkers have suggested that all of our behaviours are driven by a desire to love and be loved and to seek the joy of achievement. But, if life doesn't work out as we hoped, an unrequited love can drive us into rage when our individual sense of power and mastery is thwarted. Much of our happiness is determined by how well we can cope with these powerful but conflicting internal emotions.

How we learn to interact with other people, both those

who are close to us and to strangers, has a big impact on the kind of life we will live and ultimately the kind of life our society lives. Adam Smith, remembered mainly for his thoughts on economics, was also a keen observer of human nature and said that the desire for bettering our condition emerges with us from the womb and never leaves us until we go to the grave. From this, he developed the theory that every person striving for self-improvement generated a common good: the famous 'invisible hand' of markets where 'by pursuing his own interest, [the individual] frequently promotes that of the society more effectually than when he intends to promote it'[5]. The 'common good' is an interesting concept and I believe there is a 'something' that binds us together as a species. I prefer to think of it as a 'common imagination' and will return to that thought later. For, before you ask the question 'how did we get into this situation?' it is important to recognise that, as strange as it seems, we are all body, mind and society.

Early societies were not much more than extended families but something happened and from society sprung civilisation. The first civilisations appeared around 5,500 years ago. Focused on the basins of major rivers, civilisations (by which we mean societies with writing) comprising of cities, large public buildings and a recognisable political apparatus of state appeared more or less simultaneously in Eurasia (the Tigris/Euphrates rivers), Africa (the Nile), China (Yellow River) and India (Indus).

To supply the cities, local production needed to be organised and trade emerged between the great centres. From these cities, empires began and by 100 AD there was a chain of empires stretching from Rome via Parthia (modern day Iran) to the Kushan Empire (modern day Afghanistan and north India) to China, forming a continuous belt of civilisations that reached from the

Atlantic to the Pacific. These areas provided not only a zone for the development of trade but also for the transfer of ideas, technologies and institutions, especially religions.

Much of the early history of the world concerns the struggles of these civilisations to repel nomadic 'barbarians', a struggle they eventually lost around the fifth century. The period between the fall of the 'classical' civilisations and the rise of the kind of society we live in today is generally referred to as the 'Dark Ages', a period in which interaction between Europe and Asia, and even within the continents themselves, became much less frequent. While Europe stagnated, great civilisations rose in other parts of the world. Islam was truly dynamic at this time, while the Maya, Inca and Aztecs in the Americas and Srivijaya in South East Asia were all vibrant. Europe was not entirely cut off from these civilisations and, as we shall see shortly, the transfer of the Hindu numbering system to the West via the Crusades had an enormous impact on the way we think about things today. It was the European Renaissance, however, that formed the world we recognise now. This was a period of learning, investigation into the physical world and advance in the arts that remains unparalleled in human history to date. It was shortly followed by the Industrial Revolution and the ascendance of the city-powered European civilisation, an ascendance that continued until its transmutation into the global, interdependent, city-powered civilisation that we live in.

This civilisation is now under a greater threat than any other since history began. Not by marauding nomads or land-greedy neighbours this time, but a fundamental error we made in the foundations. The choice of fossil fuel as the source of energy for our civilisation has become the greatest danger to its survival.

Burning fossil fuel releases gases, particularly carbon

dioxide (CO_2), which interferes with the planet's ability to radiate the sun's heat energy back into space. This is compounded by human activity such as cutting down forests, reducing the planet's ability to process CO_2. Other human activities like allowing our waste to rot in landfill sites and breeding large numbers of animals as food also causes the release of other carbon-compound gases like methane, which has an even more powerful effect than CO_2. Together these gases are called the Greenhouse Gases (GHG) and the combination of our activities is causing the concentration of them in the atmosphere to rise, in turn raising the global average temperature. The consequence of these temperature rises will be the melting of ice at the poles and glaciers of the world, causing sea levels to rise by an amount that will submerge not just the Maldives and Bangladesh but the Netherlands, London, Florida and Louisiana amongst others. The planet's ability to produce sufficient food for the human population will be severely stretched, possibly overloaded. The science of this is not in doubt. Every national science academy on the planet agrees with this scenario, the arguments are around what to do about it; when and who should pay. In other words, the arguments are political.

The political world has been grappling with the problem for some years now. In 1988, the United Nations set up the Intergovernmental Panel on Climate Change (IPCC) to assess the science and the IPCC published its first report in 1990 which confirmed the danger. In 1992, at the Rio Earth Summit, the politicians decided that action needed to be taken, but it wasn't until Kyoto five years later that they agreed upon what to do and set emission reduction targets. The Kyoto agreement runs until 2012, so recognising that something needed to be in place by its expiration date, the conference at Bali in 2007 set Copenhagen 2009 as the target for agreeing the next set of binding agreements. In the event, all that they managed to agree in Copenhagen was that cuts were needed. The world would work towards keeping the rise in global temperatures to below 2 ^{0}C and some progress was made in the area of deforestation and setting up a fund to help poorer countries adapt to the effects of climate change. Meanwhile, in the real world, emissions continued to rise and by 2007 were 29% higher than 1997 (see Chart 1). Only the UK and Germany had managed to make any reductions and even then by only a single percent digit.

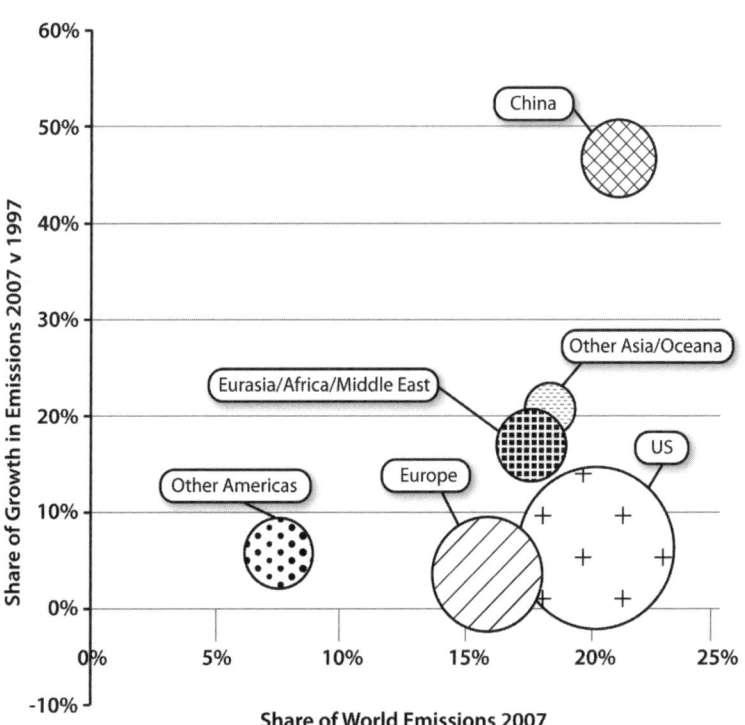

CHART 1
World Carbon Dioxide Emissions from Energy Consumption

Bubble size illustrates emissions per capita
(Source: Energy Information Administration, Population Reference Bureau, author's calculations)

If the politicians can agree that this problem is such an important issue (it is often referred to as the greatest challenge of our times), why is it so difficult to agree action? There are a number of reasons.

Firstly, the science of climate change is very complex. At one level, it is very simple: higher temperatures leads to melting ice, which in turn leads to rising sea levels (whilst that takes place, changing weather patterns will make growing crops more difficult and extreme weather events more frequent). But being able to say *exactly* when the levels will rise or *exactly* which areas will experience the most significant weather changes or *exactly* how things might change is much more difficult. It is subject to probabilities and risks.

Probabilities have a bigger impact on your daily life than you might imagine. How much you pay for your car insurance or to insure your house is directly related to probability: the assumption is that people with similar profiles face the same risks and the higher the risks, the higher the price you pay. In fact, probabilities are used by people and businesses all around the world to help them make decisions, every day.

Some probabilities and decisions are easier to calculate than others. Let's take an example: suppose I wanted to launch a career as a magician and decided that to establish my credibility, I needed a video of myself tossing a coin and 'magically' getting 10 heads in a row. Although the probability of getting a head on any one throw is 1 in 2, if I toss a coin repeatedly getting a sequence that runs head, tail, head, tail, head, tail and so on is only one out of a great number of possible outcomes, each with equal probability. By doing the maths, I know that if I flip a coin 1024 times the probability is that I will get a streak of 10 heads in a row. Let's say I can make 12 flips in a minute. That means that if I spend about an hour-and-a-half filming myself flipping a coin, I will end up with a section of about a minute where I will have 10 heads in a row and will have that 'magical' clip to put on my website. All I need to decide now is whether the time investment is worth my perceived value of the clip.

Now let's consider another example. Say I am single male in my late twenties or early thirties. I have met and fallen in love with a beautiful woman and am wondering whether I should ask her to marry me, as I'd really like to have children of my own. Should I ask her to marry me?

The decision here is much more complex, although the desired outcome is still easy to specify. How could we go about working out the probabilities and risks in this situation? In my calculations, it might help to understand how fertile the prospective partner is and how high my sperm count is, so we can work out how easy it will be to have children. It might also help to understand how compatible our personalities are, which will tell us how likely we are to enjoy living together. These calculations help but don't illuminate the key risks that I am trying to resolve: will she say yes or no? will I meet an even better candidate at sometime in the near future? The final decision will come down to my intuition.

With climate change, the decisions the politicians are being asked to make are more akin to trying to decide about making the proposal rather than making the video. There is a huge number of variables that are involved. Some can be measured, like the probability of a certain level of temperature rise given a quantity of GHGs emitted. Some are unknowable, like what will the United States do? Should we try to avert the changes we are causing to the atmosphere? Will enough countries co-operate to make our efforts worthwhile? Should we just try to cope with the impacts as they come along? Should we do a combination of both and if so, what is the best combination? Who should pay? When should we take action? These are the questions the politicians are grappling with and the scientists are telling them they need to start making these decisions *now*. Because if the politicians do not decide quickly what kind of future we want to marry ourselves to, nature is on course to make it very ugly indeed.

Politicians are dependent on the people for their power.

Even in non-democracies, the leaders need to attend to the general state of the nation if they want to avert a revolution. For centuries, politics has been practised in the same manner: the leaders of countries are concerned about the next election. In politics, humanity does not exist, only the voters. The next century does not exist, only the next year. The next generation does not exist, only the next election.

President Obama has said that he is not President of the World. He is, rather, President of the United States and must defend the interests of his country's voters, who are concerned about changing their cars and increasing their consumption, not about saving the planet. Politicians are unprepared for long-term planetary problems. They and their electorates have great difficulty in gauging what might happen in the future. When the voters in the developed West, who are the biggest emitters of the GHGs, do think about it the majority remain unmoved.

Concern about increases in expected frequency and severity of major weather events like droughts or floods is generally low in places such as the United States and Europe. This may be because low probability events tend to be underestimated in decisions based upon personal experience, unless they have recently occurred in which case they are vastly overestimated. Many think of climate change risks (and thus of the benefits of mitigating them) both as considerably uncertain and as being mostly in the future ('it is a problem for our children's children' is a fairly common view).

The risks are also considered geographically distant. The Maldives, which have a reputation for beauty and are a popular upmarket destination for well-off tourists, face being destroyed completely by rising sea levels. Whilst that may be lamented by people in Western Europe or the United States, in itself it is not a sufficient motivation for the majority who would probably never go there anyway. People are more concerned about what happens in their immediate vicinity than in far-off lands. Much more important to the voters is what is happening in their economy, now.

The two biggest emitters, China and the United States, are at very different stages of their economic development and are equally reluctant to make promises on reducing total emissions. Both accept that cuts are needed but there is an associated cost in making the change. If one country presses ahead without the other, there is the fear that their economy will suffer from higher costs of energy production without seeing any short-term advantage, so neither wants to 'go first'. Each country is waiting for the others to agree to act at the same time. It is rather like the world is engaged in a giant game of 'Prisoner's Dilemma'.

Imagine, if you will, two criminals arrested under the suspicion of having committed a crime together. However, the police do not have sufficient evidence in order to have them convicted. The two prisoners are isolated from each other, and the police visit each of them to offer a deal: whoever offers evidence against the other will be freed. If neither of them accepts the offer, they will both be charged and face court.

Now they have a choice, but making the choice depends on how they think the other person will behave.

If they both keep quiet, they can be considered to be

cooperating with each other or uniting against their common enemy, the police. They might still be charged with the crime, but there is a good chance they will be acquitted due to lack of evidence. Therefore, they will both gain. However, if one of them betrays the other one by confessing to the police, the one who breaks will gain more since he is freed; the one who remains silent, on the other hand, will receive the full punishment since he did not help the police and there is now sufficient proof with the statement of the betrayer. The silent one will face the full fury of the law.

If both betray each other, both will be punished, but less severely than if they had refused to talk as the justice system gives credit to criminals who confess to their actions.

The dilemma resides in the fact that each prisoner has a choice between only two options, but cannot make a good decision without knowing what the other one will do. This is similar to the dilemma that politicians face: everyone agrees cuts must be made, but they are frightened of putting their economies at risk. No wonder politicians prefer to talk about addressing poverty and development as priorities.

They are willing to acknowledge that climate change is the greatest threat to the future, but forgoing fossil energy driven economic growth will have to wait. After all, the energy they have provided and the technology that exploits these fuels have allowed us to thrive as a species, extending our life expectancy from 35 years prior to the Industrial Revolution to around 80 years now[6]. It also kick-started a population growth that allowed us to exist on the planet in record-breaking numbers.

Amazingly, the large majority of the world's population enjoy a life that would have been unimaginable to our forbears, although the conditions of the poorest billion would have been all too familiar. Fossil fuels impact upon every aspect of modern life (as Chart 2 shows). It seems impossible

CHART 2
UK Emissions of Greenhouse Gases 2006/7
Total Emissions = 685 million tonnes

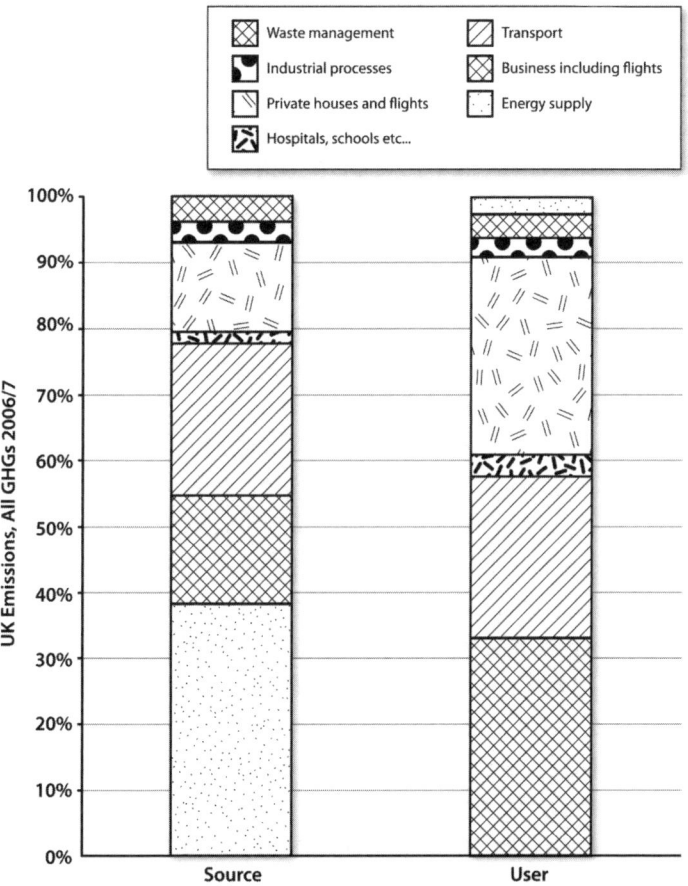

Source: Defra, DfT, author's calculation

to reduce their use without fundamentally depriving people of what they currently enjoy. If you were a politician, would you run the risk of unpopularity by making access to fossil fuels more difficult? Clearly, making significant cuts in emissions in the short-term is going to be a big task.

If politicians draw their power from the people, what can be done so that the people demand change in greater numbers? The subtle interaction between individuals and society is driven by people's behaviour. Our behaviours are driven by how we perceive the world around us and attempt to cope with it. So what is it that individuals can do to influence our common imagination toward believing that we can and will bring about the changes that the scientists are telling us are needed? We need to return to my earlier point that none of us can predict the future with precision and we each experience the world in our own way. Let's think about the future first.

Over the past 60 years we have become accustomed to defining 'normal' as change for the better, largely through the development of technology. When the Wright Brothers made the first powered flight in 1903, they could have scarcely imagined that one day thousands of people would fly around the world everyday. Nor would they have imagined that man would have 'slipped the surly bonds of earth to touch the face of God'* within just a lifetime by landing on the Moon in July 1969. But change isn't always perceived as being for the better.

My maternal grandfather fought in the trenches of World

---

* This was actually said by President Reagan, in a television broadcast to the nation on the afternoon of the Challenger disaster, paraphrasing a sonnet written by John Gillespie Magee, a young American airman killed in World War II. A similar sentiment was employed in a speech prepared for President Nixon in the event of the Apollo moon landing ending in failure.

War I, which was the only time he left Scotland. My memories of him date from the 1960s, when he was disgusted that 'pig's food' had been packaged and sold to humans as yoghurt and would leave the room when the 'corrupting' Top of the Pops, a platform for the pop groups of the day, came on television. That was not what he had risked his life for.

When we do consider the far future, we have a tendency to imagine either great disaster (an asteroid strike or some other cataclysmic event) or great improvement: the idea of a steady decline seldom makes an appearance.

When I was growing up, the 21st century seemed a very long time away and *Space 1999*, a popular science fiction programme, seemed a credible description of how we might all be living by now. It turns out that forecast was wrong. Over human history it seems we often get the predictions wrong.

Many great civilisations have come and gone. The Greek, Roman, Egyptian, Ming, Inca and Mayans were all mighty and powerful. At the time of their height, certainly, they would have seemed invincible to their inhabitants. Literature, buildings and ideas have all survived from each of them, but the civilisations themselves are long gone. The Mayans, who predicted that the world would end in 2012, were unable to foresee that their own demise would come much earlier.

Forecasting the future is uncertain and past warnings of impending ecological disaster, from Malthus onwards, have not come to pass. The idea that our civilisation might go into decline, as wealth is destroyed by extreme weather events and food becomes ever more difficult to obtain, is a difficult future to imagine. Besides, 'decline' is a dark word for humans. We may like to think in terms of giving things up, such as smoking, or cutting down on things such as alcohol.

The motivation here is usually positive and we expect a better health for our body as a result of it. But while we may decline to do something, I have never heard of anyone willingly going into decline. A synonym of decline is 'decay', a word all too resonant with our image of what awaits our bodies when death finally comes.

Contemplating our own death is the one part of the future that few of us want to engage in. Probably uniquely amongst species, we know that we are going to die. Exactly how or when is unknown, but we live virtually all of our lives with the knowledge of its temporary nature.

Whether we consciously acknowledge it or not, we seek out life-affirming experiences by putting that very life at risk. Roller-coasters are a well-known example of a pretty safe risk. Many of the sports and pastimes we pursue have a dimension of excitement that is gained through the risks involved in them.

We each have our own approach to risk. Many of us are prone to consistent and low-level risk behaviours such as smoking, drinking excessively or overeating, believing in our personal immortality while worrying about unlikely events such as being eaten by a shark if we swim in the sea. The important thing to recognise is that the perception of risk is immensely personal. An individual's perception of risk can change, however, usually through experience.

There is a, possibly apocryphal, story of a mathematician who never went to the air raid shelters during World War II. When challenged by friends concerned for his safety, he would reply: 'There are millions of people in this city, why should I think they are going to bomb me?' One night, however, he showed up in the shelter. Asked what had changed his mind, he replied: 'There are millions of people in this city but only one elephant in the zoo. Last night they got the elephant.'

This is one of the reasons why Smith's 'invisible hand' works. Different people make different assessments of the likelihood of different futures, then decide to act accordingly.

People need to lend or borrow money from each other in order to undertake economic activity. The lender believes that the future returns are worth the risk of his money disappearing. The borrower believes that the share of his future profits he will have to give to the lender is justified when set against the prospect of being unable to raise the money at all, thus being denied the opportunity to conduct his business. In other words, their view of the risks involved is balanced. However, as anyone who has tried to borrow money or watched *Dragon's Den* can testify, finding this happy situation can be extremely difficult.

In climate change terms then, the risks perceived by the scientists are simply not those perceived by the politicians or the public at large. The majority of people simply do not yet see themselves as being at risk from its impacts. If you form part of that majority, the first Labour will, I hope, get you to reconsider.

Whether you believe in Smith's 'invisible hand' theory or not, another highly personal and idiosyncratic element that affects how our society works is the concept of value.

Value can be thought of as the individual's feeling towards what he has received in exchange for what he has given in a transaction. What he has to give usually involves money, but it can also include consideration of the time and effort required to make the transaction. What he receives is not just the product or service in question, but also includes perceptions of the quantity, quality, status and convenience he feels it provides.

The difference in people's perception of value is what drives modern civilisation: people prefer to pay others to provide things rather than producing them for themselves. This has always been the basis of trade: an exchange takes place and both sides come away feeling better off after it.

When I buy something, it is because the value of it to me is higher than the money being asked for it. Similarly, if I want to sell something, I need to find somebody who places a value on the item that it higher than the money I am asking for it. But there is a stranger aspect to the concept of value, which is that people seem to value things that they already have more highly than those which they may acquire in the future.

An experiment by Professor Richard Thaler of the University of Chicago gave one group of people, the Sellers, a coffee mug and asked them at what price would they part with it. They asked another group, the Choosers, who didn't get mugs, whether they would choose to receive the money or the mug at a number of different price points. The Sellers priced the mugs at $7.12 on average, but the Choosers were only prepared to pay an average of $3.12.[7] This behaviour is in keeping with another economic principle called discounting.

Discounting is the difference between the perceived value of something now and its value at some point in the future. Generally, people prefer instant gratification or at least short-term benefit to any long-term gain. Combined with the uncertainties of exactly what the effects of climate change will be and when and where they will be felt, this is a powerful brake on action.

People will generally act to move away from something unpleasant or move towards something appealing. The current situation is that the effects of climate change are not being felt to an enormous extent, so there is no motivation to 'move away from'. The alternative vision of life put forward by most Green groups of a return to a more pastoral way of life is unattractive to the mainstream population, so there is also little motivation to 'move towards' it. Something fundamental has to change.

One of the fundamental building blocks of our civilisation is money. Money is a remarkable thing. The story of money is virtually the history of civilisation itself.

At certain points in history, the amount of money available has been restricted to the availability of some tangible thing that functions as money, such as gold. Something surprising in the history of money was revealed to me on a visit to the National Maritime Museum in Greenwich. A display on the great voyages of discovery in history included the voyages of the conquistadores. The great hoards of gold and silver that the conquistadores seized and brought back did not provide great riches for the Spanish people, but instead provoked inflation and a collapse of the Spanish economy. Despite the riches flowing from the New World, Spain went bankrupt.

Gold continued as the basis of currency right up to the early part of the 20th century and many countries and individuals still hold gold reserves as part of their store of wealth. Quite why a shiny metal that can provide neither sustenance nor shelter managed to acquire its magical properties remains a mystery. These magical properties have now been transferred to numbers in an electronic file or prettily printed pieces of paper, but money continues to have a value simply because we confer a value upon it. Money is proof positive of the power of our collective imagination.

For an imaginary construct, money exerts great power and people will do many things to get it or to avoid losing it. There is a widespread belief that getting more money will bring happiness and that losing it will bring misery. We have developed complex systems to signal how much of it we have got, but it is usually a social anathema to ever discuss the actual amounts.

One of the strangest things to get your head around is that the economic theory which we use to make big decisions about how we do things in the world is based upon the assumption that people behave rationally. Just how strange this assumption is can be illustrated by considering the symbolic nature of 'money'.

When I use money to buy something, the transaction is not merely the exchange of money in return for a thing. It also reflects my feelings about myself, how I got the money in the first place, the thing I am buying, the location I have chosen to make the purchase, the people who work in there, the people who transported the thing there, the people who made the thing and the owners of the companies that did all these tasks. In fact, it is likely that thousands of connections have brought myself and the thing together. How I process all of these complex interactions within my decision to 'buy' or 'not buy' is supposedly summed up in the economist's concept of 'rational behaviour'. In reality, irrationality is the natural human condition.

In another experiment by Professor Richard Thaler, he gave $30 each to a class of students. He then offered them the option of betting on the toss of a coin, wherein one side would gain a further $9 whilst the other would lose $9. 70% of the students accepted the challenge. At the next class, he offered them the option of taking a bet on the coin where heads would gain $21 and tails $39. If they preferred to forgo the bet they could just take $30. In this case, only 43% of the students took on the bet. In rational terms, the possible outcomes were exactly the same, the chance to end up with $21, $30 or $39, but the sequence of choices they were offered greatly changed people's reactions. In some ways, this behaviour can be explained by how people seem to value incremental increases in relation to the amount of money they already have. If you are poor, getting hold of £100 is a big thing, but if you already have a few thousand pounds in the bank, then getting an extra £100 is nice but no big deal. Already having got some money, the students in the first experiment were more willing to risk a part of it to seek an extra bonus.

Another experiment is less easy to understand. In this one, Thaler asked people how much they would be prepared to pay for the removal of a 1 in 1000 probability of immediate death and how much they would have to be paid to accept a 1 in 1000 probability of immediate death. Typically, people would not pay more than $200 to have the risk removed, but would need to be paid $50,000 in order to accept having the risk in place where previously there had been none.

It is not just our relationship with money itself that needs to be considered, but how we choose to account for the way it flows through societies.

Classical civilisations like that of the ancient Greeks had forms of money, banks and industry that would be pretty much recognisable to us today. Indeed, one of the richest and most famous of all Greeks was called Pasion. Pasion started his banking career in 394 BC as a slave in the service of two leading Athenian bankers. He rose to eclipse his masters, in the process gaining not only his freedom but also Athenian citizenship. In addition to his banking business, he owned the largest shield factory in Greece and also conducted a hiring business that lent domestic articles such as clothes, blankets and silver bowls for a lucrative fee.[8]

What the Greeks did not have was a numbering system that would allow them to calculate rather than simply record. This restricted their thinking so although they could see and record order in the skies, when it came to figuring out what is likely to happen on Earth in the future they had no tools to help them. The future was very much in the lap of the Gods. Nor did the Greeks have a form of accounting that we would recognise.

Both of these conditions changed when the rise of Christianity and the Crusades introduced the now-familiar Hindu numbering system to Europe. This had the effect of promoting the kind of abstract thought that comes through calculation and ultimately transformed human history.

The early Middle Ages and the Renaissance were critical periods in the formation of many of the norms of civilisation and trade that we still use today, but one concept developed then which has been more pivotal in shaping the way we interact today than any other: the development of double entry book-keeping which still forms the basis of accountancy today.

As a concept, it was first mentioned by Fibonacci in 1202, but was not fully formulated until 1494 when an Italian mathematician called Paccioli wrote about it in a book called *Suma de arithmetic, geometria et proportionalita*. It may seem a strange, almost outlandish thought, but double entry book-keeping revolutionised the way men conducted business and made decisions. Concepts such as profit margins, budgets, rates of return, cost-benefit analysis and virtually every other decision-making tool in common use today were simply not possible before its introduction[9].

Unfortunately, in the 15th century there were many fewer people alive and we had not begun to use fossil fuels in any sort of meaningful way. Wood was still the principal source of fuel. Moreover, the dominant world view in medieval Europe was Catholicism, which held that the world was man's dominion, as gifted by God. Under such conditions, it is unsurprising that no thought was given to understanding the 'cost' to the environment of using a product.

Economic activity was considered only in terms of the

land, capital and labour that was used to produce an object and these were the only costs that were assigned. Even when economic activity grew and the adverse effects of the consumption of fossil fuel were noted, the environment was simply treated as an 'externality' by economic theory. An 'externality' is rather a neat intellectual trick which has the effect of sweeping something under the carpet.

This accountancy error is almost 500 years old and is the core thing that needs to be fixed. There is a cost to using the Earth's environment which has been omitted and we need to change the way we think about how we account for it. The 'Earth-cost' of consumption has to be included in the price of production.

Achieving that will not be easy, as it requires overturning around 500 years of convention and habit.

Some governments have started to try to introduce the concept through the development of carbon markets, but these have not yet had any real impact on emissions. I believe we can achieve this change and in a timescale which will give us a reasonable chance of reclaiming a safe climate, as long as a sufficiently large number of individuals decide that is what they want to happen. In large numbers, we can shift our collective imagination and change the rules of accountancy. To achieve this, however, will take time and we must take steps to reduce our emissions whilst working towards a lasting solution.

Seemingly impossible tasks have been achieved throughout human history, so let us look to history to inspire us.

Our current industrial civilisation can trace its roots back to the ancient Greeks. It was they who first started to look for 'proof', to answer the question of why something was as it was, rather than simply seeking to describe it. They sought to develop concepts that would apply everywhere and in every case. The work of Pythagoras is as valid today as it was 2,500 years ago. The philosophies of Plato, Aristotle and Socrates are still read. So I will turn to the Greek myths to guide me as I consider the actions needed to challenge climate change. Myths are both stories and powerful tools for gaining insight into how things are or may be. They don't provide answers to particular situations, but they do provide ways of starting to think about answers. They help us to get to grips with understanding how we may behave differently.

The myth I have chosen is that of Heracles, or more particularly, the Labours of Heracles (or Hercules in the Roman and Disney versions. As this is the more popular spelling, and as the story is essentially the same in both Greek and Roman versions, I shall use it for the remainder of the text). The Labours are particularly apt here, not just because Hercules was reputed to have made Earth safe for mankind by completing them, but the situation before them has striking parallels to the situation we find ourselves in now.

Hercules was not born as a god, but as the offspring of an extended night of sexual activity between the god Zeus and a mortal, Alcmene. Somewhat surprisingly given that he was an all-powerful god, Zeus had thought it necessary to trick Alcmene into accepting him by impersonating her husband. Immediately, we can see something relevant: a deception in order to achieve one's own way is a frequent occurrence in modern life, especially in business. In principle, trade is a transaction in which, after it is done, both parties feel better off for having done it. In reality, success in business is often at the expense of someone else. Creating a situation in which people believe that the benefit they are going to accrue from a transaction will be greater than what they will actually get is the primary job of advertising. Retaining the maximum amount of money the company receives is the job of accountants specialising in tax avoidance.

Zeus was married to Hera, the Queen of the gods. Not content with simply cheating on her, Zeus declared that the boy was to be named in her honour. Maybe he really thought she would be pleased: after all, Zeus's plan all along had been to beget a son powerful enough to protect both gods and men from destruction. Hera wasn't at all pleased about his infidelity and wanted revenge. Men have a long history of not understanding women particularly well. Hera did, however, realise that she could not win a straight fight against Zeus, so decided instead to make life a misery for the boy.

Before taking on his Labours, Hercules had led a pretty tumultuous life. By all accounts, he was a keen and active student of chariot driving, boxing and the arts of war, especially archery. He also learnt music but when a teacher called Linus stood in for his regular one and tried to get him to do things differently, Hercules lashed out at him and killed him. This prompted his earthly stepfather to send him away to the country to be a cowherd in an attempt to prevent further crimes.

The country life is seldom as idyllic as city dwellers often fantasise and here he found a family with fifty daughters with whom he proceeded to sleep. In some versions, all in the same night; in others, one after the other. He was also presented with a choice.

One day, while watching the herd, Virtue, a tall slender woman in a simple white robe, approached Hercules from one side, whilst from the other side came Vice, a curvaceous young woman wearing makeup and a dress with a plunging neckline. Vice offered him sex, entertainment and lifelong ease, whilst Virtue offered him struggle and labour, but he would be rewarded by lifelong fame. Even in the days before reality television and *The X Factor*, the lure of lifelong fame proved too much.

For us to choose Virtue will also involve struggle, but it is simply a struggle to overcome 500 years of accumulated conditioning which tells us that the rational thing to do is to burn fossil fuels.

Vice wants us to keep burning the oil and gas and, when they run out, to turn to coal. They are convenient, cheap and a potent source of energy which is fairly easily transported to wherever it is needed. And they make life so much nicer. The cars, the warm houses, the bright lights of the cities, the graceful aeroplanes that cover enormous distances, the enormous ships that bring goods from far away but cheaper lands, the high yield farms, the shiny electronic games, phones and labour-saving devices. These are all dependent on energy and coincidentally that energy comes from fossil fuels.

Virtue insists that we find a new fuel. Energy can come from renewable resources (sun, wind, tide, plants and trees), from nuclear or hydrogen. Enough energy from the sun hits the planet in one hour to power human activity for an entire year. At the moment, we find it difficult to capture that energy and turn it into something that is as convenient, storable or transportable as fossil fuel. Finding ways to do so are hampered by the accountancy method that ignores the Earth-cost of consumption when setting the price of production.

We have reached a point of evolutionary choice. We can follow our animal instincts of greed and sloth and burn the remaining fossil fuel in the knowledge that it will fry the planet. Or we can choose to change our ways of accounting, so that the cost of developing alternatives is no longer weighed as a disadvantage when set against the easy option. Vice or Virtue? Life-shortening ease or everlasting fame?

Hercules made his choice.

When he accepted the Labours, it was in atonement for his excesses. He was well supported by the other gods, each gifting him an item that would become useful. Zeus gave him a magnificent and unbreakable shield; Poseidon a team of horses; Apollo a bow and smooth-shafted arrows. When he went into the Labours, however, Hercules needed skill, courage and innovation. He also had to complete them all. In the next section, you will find twelve Labours for you to take on to maximise your impact in the fight at both an individual and a global level. I have tried to focus on where the biggest impacts can be made and the Labours start in areas in which you have the most control and continue to areas in which you have the least. I have also tried to provide pointers to enable you to find tools and sources of help. There is even a cash gift available to you when you get to the section on powering your house.

For maximum impact, you should take action in all 12 areas. There are plenty of references to additional material if you want to debate the data, but don't let that distract you from taking action now. To be successful, we need to *act*. To be truly successful we need to act at a personal, local, national and international level.

PART TWO

The Labours

Deciding to Act
(The Labour of The Nemean Lion)

The stories of the Labours of Hercules are not exact accounts of what truly happened. Passed down through oral histories and shaped by the cultural nuances of both ancient Greece and ancient Rome, there are many variable and even disputed details. Different versions of the myth contain differing emphases, such as the motivations they attribute to those who interacted with Hercules as he undertook the Labours. But in each, there is an essential truth: a challenge and an action in response to it. And so it is with climate change. There are many versions of what might be happening now, many motivations for the actions of the actors taking part, but ultimately there is a story that is unchanging.

The First Labour of Hercules was to kill and flay the Nemean Lion, a fearsome beast that had a skin that was impervious to all materials known at the time including those used in weapons. That didn't stop Hercules from trying to tackle the lion with arrows, a sword and a club. Somehow he had to find out for himself that his weapons were useless.

For the majority of people looking at climate change, it appears that this is the phase we are stuck in, looking for 'the weapon' that we can use to defeat it.

On 30 April 2009, *The Sun* newspaper (and others) published reports that said: 'Britain set for a Barbeque Summer [...] with warmer than average temperatures and near or below-average rainfall'. By 29 July, *The Times* (and others) carried reports which said: 'The Met Office officials refused to apologise today after admitting that the "barbecue summer" they had predicted was no longer likely'. To the Great British public, global warming and climate change took another back seat.

The primary difficulty with climate science is that it is incredibly complex and complex systems are still very difficult for our species to understand, even for those who are supposed to be experts. This was clearly demonstrated in 2008 when the financial system collapsed, causing a deep economic recession.

The first half of 2009 saw a 50% reduction in production output of automobiles[10] and the overall economic output for the 12 months up until mid-2009 shrank by 5.6%.[11] The root cause of the crisis is generally agreed to have stemmed from the high levels of risk that were taken by the banking system in lending to the 'sub-prime market', that is to people who would not previously have qualified for loans by virtue of their probable inability to repay them. When these borrowers started to default, the loans had been incorporated into other financial instruments to such an extent that nobody could figure out how big the problem was or who was most exposed to the potential losses. Confidence collapsed. Banks stopped lending to each other. The flow of money was greatly contracted.

The scale of the ensuing crisis prompted the Queen to enquire of economists why no one had seen it coming. In response, Tim Besley of the London School of Economics and Peter Hennessey, an eminent historian of government, submitted a letter which concluded:

'In summary, Your Majesty, the failure to foresee the timing, extent and severity of the crisis and to head it off, while it had many causes, was principally a failure of the collective imagination of many bright people, both in this country and internationally, to understand the risks to the system as a whole.'

What we do know about the climate system, and have done since 1824, is that the planet is affected by a greenhouse effect, where some of its radiated heat is reflected back from the atmosphere to the surface. This effect makes the planet warmer than it would otherwise be expected to be if it were simply receiving and reflecting the sun's energy.

We also know that the strength of this greenhouse effect is made stronger by the concentration of some gases in the atmosphere, notably carbon dioxide (CO_2), methane, nitrous oxide and halogens also known as CFCs. Collectively these gases are referred to as green house gases or GHG for short.* For most of human history, the concentrations of these gases, especially CO_2, has fluctuated between 200 and 280 parts per million (ppm). Since the Industrial Revolution, they have risen sharply and the rate of increase in recent years has gone up so they are now about a third higher than the long-term average.[12]

---

* These latter chemicals were formerly used in refrigerators and as propellants for aerosols, and their presence in the atmosphere was identified as the cause of a hole appearing in the Earth's ozone layer, which protects the Earth's surface from some of the more dangerous wavelengths of the sun's radiation. Their use was banned in an international treaty signed in Montreal in 1987 and their concentration in the atmosphere has now started to decline.

Many human activities use power generated from burning fossil fuel and this has been the primary source of the increases.

Using coal, oil and gas to generate our electricity and petrol, along with diesel and kerosene to power our transport, has dumped billions of tonnes of the carbon dioxide formerly trapped as fossil carbon back into the atmosphere. Cutting down trees and slash-and-burn clearances to gain more farmland has added to the problem, but the fossil fuels have created nearly three and a half times as much. Breeding animals to feed ourselves and dumping our waste to rot in out-of-sight pits has generated millions of tonnes of methane, another carbon compound and one whose ability to reflect heat back to the surface is nearly 25 times more potent than CO_2, but it still accounts for only about a quarter of the impact that the fossil fuels do. The use of oil-based fertilizers releases another greenhouse gas, nitrous oxide, which many of us have experienced as the 'laughing gas' used in anaesthetics at dentists and hospitals. Nitrous oxide is nearly 300 times more potent than CO_2 but thankfully the quantities in the atmosphere are much smaller and it makes up less than 10% of the greenhouse gasses even when its additional potency is taken to account[13]. The single biggest source of increased concentrations is that remarkably cheap, highly concentrated and fairly convenient group of fuels; coal, oil and gas.

Because there is a higher concentration of greenhouse gases in the atmosphere, more heat is reflected back to the surface. More heat at surface level causes the polar ice to melt and promotes a greater evaporation of water, both of which affect the climate of the planet and the levels of the seas. This part of the physics of climate change doesn't seem to be challenged by anyone. It is a recorded fact that

CHART 3
Concentration of CO2 in atmosphere in parts per million (ppm), past 1000 years. February 2010 reading = 390ppm

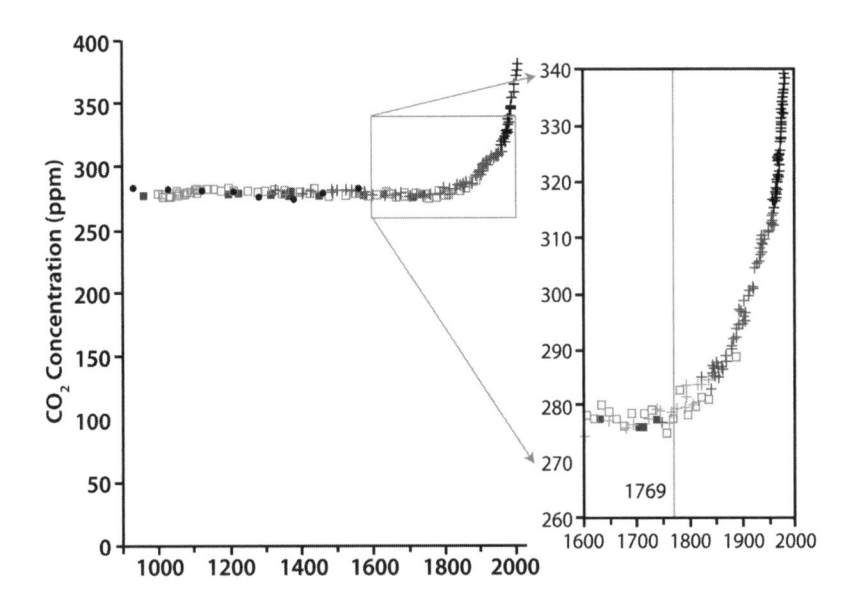

Source: David J.C. MacKay, Sustainable Energy – without the hot air (UIT, Cambridge, 2008), NOAA/ESRL

global average temperatures have increased by 0.8° C from pre-Industrial Revolution levels. What is being debated is how much more temperatures will rise, how and when the climate will change and what we should do about it.

For those who do engage with the issue, there remains the problem of the numbers. Greenhouse gases are counted in parts per million and the changes in temperatures are expected to be just a few degrees. Can there really be anything to worry about? Why aren't the numbers bigger? There are two things here that we need to recognise. Numbers are an invention of our civilisation. The Piraha, an Amazonian tribe, manages quite well without them[14]. Most people get used to dealing with a certain magnitude of number and have great difficulty noticing potential consequences when magnitudes change abruptly.

At the start of the credit crunch in 2008, one of the biggest public outcries came after the size of the pension of the Royal Bank of Scotland's departing Chief Executive was revealed. This was at a time when his bank (one of the worst affected) and indeed the whole banking sector had suffered enormous losses that ran into tens if not hundreds of billions of pounds. John Lanchester in *Whoops!*[15] speculates:

'Why did the government go to such lengths to secure Sir Fred's acceptance of his own departure, rather than just sacking him? Why did they agree to the doubling of his pension pot? We don't know, but it's almost certainly because, when the deal was done, everyone was so preoccupied by the question of whether the British banks would stay solvent that Sir Fred's pension was the last thing on anybody's mind.'

So, the £16 *million* was not a big enough figure to really be noticed when all the other discussions were about the £25.5 *billion* the government had injected into the bank, or the £302 *billion* of assets that were put into the government's Asset Protection Scheme. Even if someone had called Sir Fred to account, he would probably have pointed to the £17.7m pension pot of Stephen Green of HSBC Holdings and claimed he was simply asking to be treated fairly among his peers.

The other point about the numbers is that parts per million is a scale and many people have difficulty relating to the differing impact scale values can have. Think about speed, which is a scale in Britain we normally measure in miles per hour. Our 'natural' walking speed is about 2 to 4 mph. This speed can kill us if we fall over but it is pretty rare and anyone meeting their end this way would generally be considered to be unlucky. We have managed to increase our speed, first by using animals then by using powered vehicles. The effect of being hit by one of these powered vehicles is significant and strongly related to the speed it is travelling at. For every 20 pedestrians hit by a vehicle travelling at 20 mph, one will die. If the speed increases to 30 mph, nine of them will die. Raise it again to 40mph and the death rate goes up to nineteen out of the twenty dying[16]. By just doubling the speed, you have completely flipped the chances of survival. These are similar to the risks we are dealing with in the parts per million of greenhouse gases.

In addition to the difficulty of dealing with the numbers, there is the social argument about whether the rising levels of greenhouse gas have been caused by human activity or are just part of a natural cycle. In reality, the scientific community has stopped arguing about this. The US National Academy of Science warned of the climate impact of human activity way back in 1979 and since 2007 there is not a single scientific body of national or international standing that has maintained a dissenting opinion.

This does not mean that there are no dissenting voices. But they are from individual scientists or from bodies with scientific sounding names that are funded by the fossil fuel industry to ensure that 'doubt' remains a powerful disincentive to action. In his book *Heat*, George Monbiot dedicates an entire chapter to what he describes as 'The Denial Industry' and lists a number of fossil fuel funded organisations that make it their mission to 'take a consistent line on climate change: that the scientist is contradictory, the scientists are split, environmentalists are charlatans, liars or lunatics. And if the government took action to prevent global warming they would be endangering the global economy for no good reason. The findings those organisations dislike are labelled "junk science". The findings they welcome are labelled "sound science".'[17]

At the heart of the denial is money and self-interest.

Economics is not a science in the strictest sense, in so far as economists do not undertake experiments to test hypotheses, the results of which would be peer reviewed before a new understanding emerges. Economics is, rather, a social science which tries to understand how people make choices between different courses of action, particularly where the choices are finite, e.g. it is either 'do A *or* B' and cannot be 'do A *and* B'.

Central to the workings of economics is the concept of scarcity, the idea that people want more than is available. Scarcity limits us both as individuals and as a society. As individuals, limited income (and time and ability) keep us from doing and having all that we might like. As a society, limited resources (such as manpower, machinery, and natural resources) fix a maximum on the amount of goods and services that can be produced.

The mechanism that enables choices to be made is money and acquiring and using money takes up an enormous amount of our time and thinking capacity. It also has a pretty strong influence over our behaviour in many situations. This is quite surprising in some respects because money itself is a concept, which sometimes manifests itself physically in a coin or printed piece of paper, but more often as a balance in a ledger somewhere. The simple solution to the scarcity of money would therefore appear to be to just make your own money, but over the years societies have developed a complex set of rules over what are acceptable and what are unacceptable treatments of money in order to regulate its value. So, for example, an individual who prints their own money, no matter how identical to those notes printed by the government, is treated as a criminal and usually jailed if caught.

Catching them, however, is obviously something of a problem: a sample of coins in circulation made by the Royal Mint in 2008 found that 1 in 50 pound coins was a fake, with the total value of fakes worth about £30 million[18]. That confidence in the currency has not collapsed by this development is probably due to the fairly limited value of a single pound. Falling foul of getting one in your change is a fairly minor inconvenience and anyway the 30 million false coins are a tiny proportion of the estimated 44,900 million pounds in notes and coins in circulation in the UK.[19]

Global concern that 'global warming is a serious problem' is consistently high, with 75% of people in the US in agreement with that statement, with a similar level in Russia (73%), which are both lower than the percentages in many other nations. 87% of Canadians, 81% of Mexicans, 95% of French, 88% of Chinese, 97% of Japanese, 96% of Brazilians, and 94% of Indians assess global warming as a 'very' or 'somewhat' serious problem[20]. What seems to be lacking is the capability of taking purposeful action *now*, which is what the scientists are saying we need to do.

My belief is that this inaction is being driven by the idea that the changes are going to be small and take a long time to happen. As a Scot who has spent about half my life living in Scotland and the other half living predominantly in the southeast of England, I can attest that a small difference in average temperatures does make a noticeable and agreeable difference to general climate (Glasgow is about 1.5° C cooler than London). But the levels of temperature change that are being bandied about (not just as probable but, by some, as acceptable) are in the order of 2 to 4° C as a global average and up to 10° C at the Poles.

The Copenhagen Accord did not require countries to

make pledges on emission cuts, but according to most estimates, the proposals that were on the table but not formalised would likely have led to a 3°C rise.

The Hadley Centre has published a series of forecast maps of what this would mean for different regions of the world and all of them are pretty scary. But still the media and the Internet carries dissenting voices that deny that climate change exists. It doesn't matter. On the basis that I do not think you want to spend 10 years studying climatology in order to decide who is 'right', I would like to invite you to take a small thought experiment. When global average temperatures were of the order of 3°C cooler than the time of the Industrial Revolution, the Earth was experiencing the last Ice Age and the ice sheet extended south of London and New York. Since the Ice Age, the temperature has risen by 3.8° C so how do you think the Earth will look after a further 3°C?

As for the timing, I invite you to make your own predictions. Have a look at the chart below and see when you think the Arctic will be entirely clear of summer ice. Do you think we can wait?

Global warming and the resultant climate change is a real phenomenon.

The primary driver is the total amount of excess carbon dioxide entering the atmosphere through our choice of using fossil fuel. This is a fuel problem rather than an energy problem.

An accounting error introduced by an unfortunate sequence of events has led to powerful vested interests and a social and economic system that combine to encourage the continued use of the wrong fuel.

The world's use of fossil fuel is large, but it is not the 'be all and end all' of the global economy. Complete data records for coal, oil and gas markets is difficult to acquire,

CHART 4
Arctic Sea Ice Loss

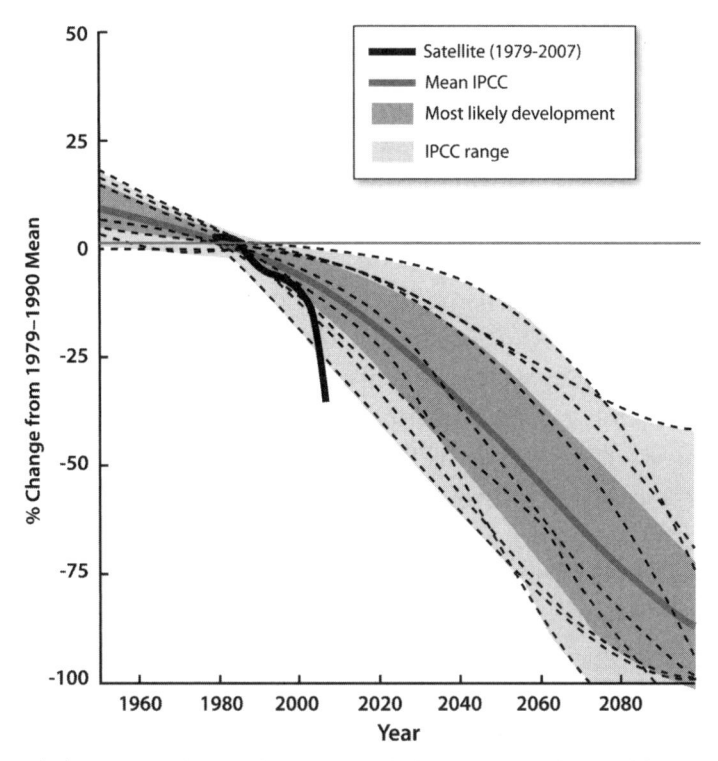

Arctic ice extent loss to September 2007 compared to IPCC modelled changes.
Chart courtesy of Dr Asgeir Sorteberg, Bjeknes Centre for Climate Research and University at Svalbard, Norway.
Date 23 September 2007
www.carbonequity.info/images/seaice07.jpg

but the value of crude oil used globally in 2009 is estimated at 3.5% of the world's income. Allow for value to be added through processing and taxation costs, then add an estimate for coal and gas, and you still will be under 10% of global income.

Remember that it is energy that supports civilisation, not fuel. To put a safe climate at risk for the sake of a sector that is at best a tenth of the world economy seems insane.

The impacts of climate change such as increased storms, rising sea levels and reduced ability to grow food have the potential to decimate world wealth and maybe even the human population. The cost of Hurricane Katrina, which hit New Orleans and surrounding areas in 2005, was $1.4 billion in regard to damage *to the wastewater utilities alone*. The floods in Cumbria in 2009 are expected to cost insurance companies £100 million. Families that didn't have insurance are left to face the costs of cleaning up, which can easily exceed £20,000, on their own. Eventually, the insurance industry will not be able to cope with these losses and the all the cost will be borne in full by individuals.

How we respond to all this is a political issue, which is to say, an emotional one. When Hercules realised that his weapons were no use against the Nemean Lion, he trapped it in his cave and wrestled with it until he overcame it. Having conquered the lion, he was then still faced with the challenge of how to flay it and it wasn't until he had the brainwave of using the beast's own claws to cut through its skin that he was able to complete the task. Having done so, he then used the skin to protect him during some of the further Labours he had to undertake. This is what we must do with climate change. Standing back and waiting for others to act, or promising to change our ways if they will, are the weapons of mortals. Superheroes make the choice to get involved. By acting with authenticity and immediacy to reduce our own use of fossil fuel, we can demand of our politicians that they do the same on a national and global scale. By believing in the power of our Labours, we acquire a cloak of unequalled strength to protect us in the challenges ahead.

Transport
(The Labour of The Stymphalian Birds)

The choice of birds to represent the transport issue might at first sight seem like a reference to air transport but it is not. The Stymphalian birds were involved in one of Hercules' most difficult tasks. The birds themselves were both man-eaters and had dung that was poisonous and blighted crops. Extremely numerous, they lived on a marsh that was neither firm enough to walk upon nor wet enough to use a boat to cross. Eventually, the only way Hercules was able to get rid of them was to scare them off using a pair of castanets supplied by Athene. Once they were airborne, he picked them off with his arrows. He may not have killed them all, but he unsettled them enough to cause them to flee permanently.

According to current evolutionary theory, modern humans emerged in Africa and spread out from there about 50 to 70,000 years ago[21]. Our histories are primarily stories of war or great journeys. Marco Polo and the Silk Route; the voyages of Magellan; de Gama and Columbus; James Cook, David Livingstone and Scott of the Antarctic; the Montgolfier brothers, Louis Bleriot and Charles Lindberg; Neil Armstrong, Buzz Aldrin and Michael Collins. Travel, if not in our DNA, is certainly in our blood. To tell people not to travel is patently ridiculous. Just like the challenge provided by the Stymphalian birds, we cannot hope to 'defeat' the opponent but must find an alternative way to 'win' during the course of this Labour.

Throughout history, humankind has sought to travel further, faster and more safely. Early vehicles and vessels used animal or wind power. The first fossil fuel powered vehicle was produced in 1769, the first gasoline car 124 years later. The first production model appeared just 3 years after that when a company called Duryeas produced 13 copies from the same design.[22], Things really got into their swing, though, in 1908 when Ford started to produce the Model T. There are now an estimated 850 million cars on the planet, about one for every seven people. There is a seemingly insatiable thirst for more. The cheapest available, the Indian Tata Nano costs around £1,400, but you can spend 100 times more than that if you wish.

In Britain, three quarters of households have access to at least one car and each car in the nation's fleet travels an average of around 8,800 miles per year. Owning a car is seen by most as an integral part of modern life and the type of car you own speaks volumes about you. In 1986, the then Prime Minister Margaret Thatcher said: 'A man who, beyond the age of 26, finds himself on a bus can count himself as a failure.'

Driving a car gives the user a sense of freedom, success, power and individuality. The advertising of cars is connected with positive symbols and many people actually come to believe that their car expresses these positive feelings and symbols for them.[23] So much so that politicians, journalists and marketers use car brand names as a shorthand to describe grouping of individual: 'White Van Man', 'Mondeo Man' and 'Volvo Driver'. Despite, or possibly because of this, the status of the car is generally higher than other forms of transport, even though they can feel like mobile prisons when the occupants are stuck in traffic jams.

They are also one of the worst sources of carbon dioxide pollution.

Passenger cars produce about the same amount of CO_2 as heating our houses, but the amount of time we spend in our houses is far greater for most than the time we spend in our cars. Cars are therefore probably your single most intense use of carbon, unless you fly more than a couple of times a year.*

Given our collective desire to enjoy the freedoms and status of car ownership, it might, at first glance, seem an impossible task to cut the carbon output from personal transportation without sacrificing a hugely important part of our lifestyle. Surprisingly, there are an amazing variety of options available to you if you choose to do so.

The first one is to leave your car at home. In 2008, the average car-driver made 410 trips, of which 90 were less than 2 miles.[24] Using a bicycle for these trips would be a simple step and most of us would be able to cycle 2 miles without sweating so much that we would need to shower.

For journeys between 2 and 10 miles, bikes are still great modes of transport and, if you are in an urban area, a door-to-door trip of around 10 miles would compare favourably with public transport in terms of journey times. Depending on how quickly you cycle, you might work up a bit of sweat travelling these distances, so showering and a change of clothes might become a consideration. The number of trips by car of this length is 224, suggesting it is a common commuting distance. If that is true for you, you could perhaps ask your employer to enrol in the 'Cycle to Work' scheme. This allows you to buy a new bike at a discounted rate and the payment is deducted from your gross income

---

* The calculation of your carbon footprint is a complex task and different forms of calculation will give you different answers as they make different assumptions. It will also vary by whether you measure your own total as an individual or as part of a household. It is worth doing though and there are plenty of tools online for you to experiment with.

over a period of 12 months, which also means your employer will pay slightly less in National Insurance contributions on your earnings over the period. This should more than compensate them having to front up the cash for the bike in the first place. Have a look at www.cyclescheme.co.uk for details.

If a bike doesn't appeal to you, the next simplest solution is for you to change your car to an electric or hydrogen-powered alternative. Provided your electricity supply is coming from a renewable source, electric vehicles are a perfectly sensible step for sustainability and, even using the current grid source, are about 70% greener than fossil powered ones. The typical arguments against them concern their limited range, low speeds and high capital costs. The range argument doesn't really hold up. A typical electric car will have a range of 60 miles. As more than 90% of the journeys undertaken in the UK are under 25 miles in distance, a single charge would provide for most of the return journeys people make. With regard to speed, the new US Tesla or UK Lightening models can accelerate from 0-60 mph in 3.9 seconds and have a top speed of 125mph. That is the kind of performance that is usually associated with so-called 'super cars' such as Porsche and Lamborghini. Finally, let's turn to cost. Unsurprisingly, the Tesla is expensive but so are the Porsche and the Lamborghini. The smaller 'run-around' types of electric car are priced starting at £8,000, so are comparable to Fords, Renaults and Toyotas. Running an electric car frees you of having to pay Vehicle Excise Duty, the London Congestion charge (if that is where you live) and, in many places, parking charges. The cost of charging them is generally a fraction of the cost of running a petrol car.* 'Well, that's

*1.5 to 2.5 pence per mile compared with 12 to 16 pence per mile for a petrol car at 2009 prices.

great' you may say, but what about the batteries? Surely they do a lot more environmental damage? It is true that the disposal of batteries is an environmental problem, but so is the disposal of an internal combustion engine. There are no conclusive studies available yet about whether one is any worse than the other, but indications are that they are about the same.

From a logical perspective, then, the challenge to move to electric cars doesn't seem too intimidating.

There does remain, however, that 7% of journeys which are more than 25 miles. On average, these trips would take place on a roughly fortnightly basis[25]. So if you are this mythical 'average' car-user and decide to change to an electric vehicle for short journeys, you will either need to have a second car (and almost a third of households already have two or more) or be prepared to visit the car rental company fairly frequently.

An alternative to electric cars is hydrogen-powered cars. These are commercially available in California now for a lease rate of US$600 per month for a Honda FCX Clarity sedan. The big advantage they currently have over electric cars is their higher range of 240 miles and their refuel time and method is similar to petrol or diesel cars. The drawback is hydrogen production and distribution. Hydrogen needs energy for its production, although it is well within the capabilities of concentrated solar power to supply that. The distribution system is presently small so will need to be developed in order encourage uptake. However, given the attractions of this substitute technology, there will almost certainly be companies willing to invest in it.

Of course the uptake of new technologies is often dependent upon a new infrastructure being available to support it. Big changes can come quickly if there is a high expectation of profit (e.g. the growth of the internet) or a desired social benefit (e.g. making more university places available.) Government policy is a hugely powerful tool. Politicians are reluctant to make policy decisions that they believe will be unpopular, so whilst they are currently exhorting people to save fuel and cut down on travel, there is still a road-building programme designed to make it easier to continue travelling by car. Given the love affair humans seem to have with their cars, I would like to suggest to them a policy that could rapidly transform personal transportation and may even become a popular choice. Why not announce that in 3 years' time there will be a strictly enforced national speed limit of 40mph for all fossil fuelled vehicles but that there will be no limit applied to electric or hydrogen ones?

If the prospect of trading in your old car for an electric or hydrogen one seems a bit too daunting under current conditions, but you are planning to update your car anyway, visit www.whatgreencar.com for guidance. They have developed a rating system that expresses a vehicle's lifecycle environmental impact as a score out of 100, ranging from 0 for the greenest vehicles to 100 for the most polluting. It is a comprehensive scheme that assesses the environmental impacts associated with a car's use and manufacture, rather than simply the CO_2 emission per kilometre. This includes all aspects of producing and using the fuel: primary production, extraction, transportation and refining as well as the vehicle's consumption, and it also includes the impact of the vehicle's manufacture, assembly and disposal. It is an independent service and easy to use. On their scale, the G-whiz electric car scores best with 6 and the Hummer H3 a massive 91. The scores, as you can image, pretty much reflect the car's size but there are some surprises too: a BMW 3 series scores 39 whilst a Saab 9-3 gets a 47.

Chances are, however, that you are not going to buy a new car this year as only about one in 20 cars on the road are in their first year of registration. Whether it is new or old, if you are a car owner, the key to your carbon intensity is the number of miles you travel in it and the number of people who travel with you. The more miles you travel the more the total carbon dioxide emitted. Although this might seem self-evident, many people believe that the 'school run' must be a major source of emissions because of the noticeable decease in traffic in urban areas when the school holidays start and subsequent increase when they return. In fact, the school run only accounts for 4% of passenger car emissions.

There are two strategies to make a quick impact on your car emissions. The first is taking a passenger. The single biggest source of transport CO_2 is from those journeys of 5 to 25 miles producing 43% of car emissions, while the 7% of journeys that are over 25 miles produce a whopping 38% of car emissions.

The shorter journeys are probably mostly commuting, the single biggest source of traffic by type of journey taken. Commuting involves a high proportion of single occupancy vehicles. By definition, commuting is going from home to your place of work, so by deduction, your place of work may contain co-workers who make a similar journey to you. This is the type of behaviour change that people generally find difficult to engage with. They fiercely defend their, usually imaginary, ability to leave the house and office at exactly the time of their choosing, to select the perfect temperature in the cabin of the car and to relax to their favourite radio station or CD. The reality of the trip is seldom like that, of course, so why not expand your humanity and bring a buddy? The conversation can distract from the jams. If your office is quite big and you don't know everybody, then encourage them to sign up to www.shareajourney.org.uk which acts as a buddy-finder, or, if you want to widen your net beyond just your own organisation, try www.liftshare.org which will help you find either a regular commuter or someone who wants to share a longer trip.

Obviously, the arguments for lift sharing are even stronger on longer journeys and as most people will be doing them less frequently you will probably end up meeting a wider variety of people so you can view it as a networking opportunity.

Sharing your car has a multiplicative effect in cutting carbon.

If instead of becoming a carbon reduction Superhero, you just want to 'do your bit', then the very least you need to do is watch your speed. The majority of cars on motorways are travelling at 70mph or more[26].

My own driving habits were consistent with the majority, but recently I had a business meeting in Devon and decided to try an experiment. I would drive there in my normal way (trying to keep up with the traffic flow or at least the guy two cars in front of me) and I would drive back limiting my speed to either the speed limit or 60mph where the 70mph speed limit applied. I used the car's trip computer to measure the results and they staggered me.

The journey was 159 miles and the outbound leg had an average speed of 51.7mph and fuel consumption of 38.1 mpg. The return leg gave an average speed of 48.6 mph and 48.6 mpg (yes, they are exactly the same number but I did double check the numbers.) That is an incredible 27% increase in fuel efficiency for just a 6.5% increase in time travelled. This seems like quite a sizeable figure and as more than a quarter of car miles are driven on motorways or trunk roads then easing up on the speed would appear to be 'doing our bit'. By my calculation, the speed reduction would reduce car emissions by about 6%. Adding in the saving from doubling up on the commute would bring the total emission saving up to about 20%. Switching to electric cars recharged by electricity from renewable resources gets rid of them all. Sometimes to be a Superhero, you simply need to take bolder steps.

There is, of course, one form of transport that we are only too happy to share, the aeroplane. Flying somewhere is probably the sexiest, most exciting thing most people can imagine. The annual holiday is usually the most anticipated event of the year. Being flown somewhere on business by your company is confirmation of how much they value your services.

Building new airports is high on the agenda for developing countries and expanding existing ones in developed countries is seen as vital to 'maintain competiveness'.

It is also remarkably cheap. Aviation fuel is protected from being taxed by an international agreement, which means that on a cost per mile basis, flying is cheaper than virtually any other form of transport. It is now possible to fly from London to Dublin for less than the cost of the taxi from Dublin airport to the city centre. Flying to cities in the east and north of Europe has become cheap enough to allow groups of people to go for the weekend for stag or hen parties. A growth in flying is predicted by virtually every government in the world, with the UK volume expected to double by 2030, which will lead them to provide more airports so that this growth can be achieved!

Flying, however, is the single most intensive source of carbon dioxide produced by all human activities. There may be more GHG emissions from car transport in total but in terms of the number of people participating in the activity, flying is the largest polluter[27]. Not because the engines are less efficient than other fossil fuel engines, but simply because of the enormous distances aeroplanes allow people to travel. In addition to the enormous amounts of fuel they use, the fact that aeroplanes emit their exhausts high in the atmosphere means that they are more damaging than those emitted at ground level. This is a contentious view and people will try to reframe it in a number of ways. Consider this from the HowStuffWorks website:

'According to Boeing's website, the 747 burns approximately **5 gallons of fuel per mile** (12 litres per kilometre).'

This sounds like a tremendously poor miles-per-gallon rating! But consider that a 747 can carry as many as 568 people. Let's call it 500 people to take into account the fact that not all seats on most flights are occupied. A 747 uses 5 gallons of fuel to transport 500 people 1 mile. That means the plane is burning 0.01 gallons per person per mile. In other words, the plane is getting 100 miles per gallon per person! The typical car gets about 25 miles per gallon, so the 747 is much better than a car carrying one person, and would even compare favourably if there were four people in the car. Not bad when you consider that the 747 is flying at 550 miles per hour (900 km/h)!

What it doesn't mention is that the 747 carries enough fuel for a single flight to cover the miles that the typical UK car travels in a year. The greenhouse effect of just a single return flight from the UK to the US, or about three return flights around Europe, is likely to be equivalent to ALL of your other activities throughout the entire year.

Just how much of global greenhouse gas emissions aviation is responsible for is a contested figure and I have seen it range from 1.6% to near 10% with the differences mostly accounted for by the differing assumptions used about the 'forcing effect' (i.e. the additional damage aircraft exhaust causes because it is emitted high in the atmosphere rather than at ground level).

Whatever the true figure, it is a big contribution from what is a minority activity and even for that minority, it is only a minority of their time that is spent on a plane. In the UK, for example, more than half of the population took no flights in the past 12 months and the 'frequent flyers' taking three or more per year number just 11%.[28] The average American flies about twice as often as the average European, but they in turn fly 10 times more than the average Asian. Overall that means that maybe 2-4% of the world's population is responsible for up to 10% of emissions from an activity they spend about a day a year doing! Anyway you look at it, flying is environmentally very expensive.

Apart from the sheer carbon intensity, there are two other big problems with flying. The first is that there is no practical substitute technology on the horizon. Jet engines might become slightly more fuel efficient but any increase will be limited to a few percentage points. The second issue is the forecast in growth will quickly eat up any efficiency gains.

This causes even quite clever people great difficulties with reconciling what they know they should do with what they would like to do. Consider Tony Blair when discussing climate change on 30 October 2006:

'Unless we act now, not some time distant but now, these consequences, disastrous as they are, will be irreversible. So there is nothing more serious, more urgent or more demanding of leadership.'

Just two months later, Tony Blair was asked if he should show leadership by not flying to Barbados for holidays[29]: His response? 'a bit impractical actually'.

Or consider the Mayor of London's support, expressed during a visit to New York in September 2009, for a BA campaign called 'Face to Face', which seeks to get more people flying. When asked how this fitted with the Conservative Party's call for a reduction in unnecessary air travel, his spokesman said the mayor supported video conferencing as a tool for business but was keen to bring tourists and business leaders to London.[30]

The aviation issue is unlikely to be solved in the short-term by economic levers. It is simply perceived as too populist an issue for any democratic government to tackle in this way. Taxing aviation fuel at the same rate as car fuel in the UK would add about £260 to the price of a London to New York ticket[31] whilst taxing it at the carbon price rate suggested by economist Nicolas Stern would add about £130. This compares with an available return fare including taxes of £216[32].

If the political will and economic incentive is not present to solve the issue, is it possible that the public will fall out of love with the aeroplane? Of course it is. People are quite capable of changing their beliefs, even if they are deeply held; for example, a Christian becoming a Muslim or vice versa.

Quite how we generate or choose our beliefs is a complex subject, but we are often influenced by those people we look up to. So here is another thought experiment. Virgin Atlantic celebrated its 25th anniversary in 2009. The image of Virgin is fun and sexy. Customer service is a high priority and the founder, Sir Richard Branson, is a high profile and well-respected entrepreneur who many people would love to emulate or at least meet. What do you think would be the effect of Sir Richard saying to the world something like this: 'You know, it has been a great party and we love flying. But, we have spent $500 million trying to develop bio-fuels and it still hasn't worked. The carbon cost of flying is simply too great, so I am closing down Virgin Atlantic airlines and opening a fleet of sailing clippers instead'?

Or, at the other extreme, imagine Michael O'Leary, CEO of Ryanair, saying something similar. Ryanair has developed a controversial business model by flying people between mostly under-utilised airports at what appears to be very low cost and frequently advertising flights for 1 penny.[33] Michael O'Leary himself has been a larger-than-life figure and has voiced some pretty outspoken views on environmentalism and climate change.[34] What do you think would happen if Michael was to say: 'You know, we've made a pretty penny out of taking people to places they never knew they wanted to go. But I can see how the tax breaks have allowed us to offer unfair temptation and I sense a change in what people want. So I am closing Ryanair and opening a low-cost express coach service'?

Recycling
(The Labour of The Augeian Stables)

The stables of King Augeias were vast and had not been cleaned out for many years. Hercules was challenged to clean them in a single day. This he achieved by diverting two rivers so that they washed through the yard, clearing away the accumulated dung. However, while Hercules actually restored the local pastures to productive health, our continued habits of 'washing away' our rubbish into landfill sites is causing great problems as the rubbish materials give off methane gas as they decay. Whilst this gas can be captured to generate energy in some circumstances, in the main it is not. As Hercules came up with an innovative solution to overcome the challenge of limited time, we need to come up with an innovative solution to overcome the challenge of limited resources, because in many respects human progress since the Industrial Revolution has been one big design error.

The current story of human technology is a product of 'cradle to grave' design. This 'cradle to grave' flow relies on brute force (including fossil fuels and large amounts of powerful chemicals). We pull resources from the earth, shape them into a product, use it and throw it away. And we do this in quantities so enormous, it is simply unsustainable. If everyone on the planet consumed at the rate of a North American consumer, we would need the resources of five planets to supply them. At the slightly more modest rate of a European consumer, we would need three planets (if you would like to see your own impact, visit www.bioregional.com).

There is an emerging design movement called 'Cradle to Cradle', which works on principles that celebrate natural, economic and cultural abundance. It revolves around the idea that, in nature, waste equals food. All products are seen as nutrients within biological (natural) or industrial (technical) metabolisms.[35] However, it is probably going to be a while before it catches on in a substantial way, particularly as the authors seem keen to develop the movement in an overly controlling manner.

In the meantime, we have an enormous mountain of waste to deal with. It might surprise you to learn that household waste is less than 10% of the total waste produced. There is an enormous amount of waste (61%) produced by construction, demolition, mining and quarrying. In official terminology, there is also a difference between waste recycling and waste recovery. Recovery includes recycling, composting, incineration with energy recovery, plus what is termed 'Refuse Derived Fuel' manufacture which typically turns used chip fat into bio-diesel for vehicles. So when rubbish is dumped into a landfill site and the methane produced as it rots gets captured and is burned to generate electricity that counts as 'recovery'.

In some respects, recycling has been something of a success story in the UK and illustrates how impactful changes to accounting systems can be. From virtually no waste being recycled at the start of the 1990s, 34% of household waste is now recycled or composted, as is 45% of commercial waste.[36] There are, however, enormous variables in the rates achieved by different local authorities around the country, with public apathy being cited as the main cause in areas where rates are low. To some extent, this is understandable as there is very little easily-accessible information on what the real impact of recycling actually is.

The beer brand Stella Artois ran a poster campaign to let people know that they are using recycled aluminium in cans, recycled glass in bottles and recycled paper and card in packaging, so they clearly believe it is a good thing. But there is a debate around waste and resource management as to the extent to which the recycling of materials offers genuine benefits to the environment.

Often, critics of recycling say that the act of recycling may in fact have little or no benefit to the environment, suggesting that more energy may be used in getting materials to the recycling facility than is saved by the process of recycling.

Thankfully, that is not true.

In 2006, the UK's Waste and Resource Action Programme commissioned research[37] to review international studies of the environmental impact of recycling versus landfill or incineration. They used what is called Life Cycle Analysis, so looked at the impact and energy usage in all the stages from virgin material to disposal. In total, they studied nearly 200 scenarios for the treatment of glass, wood, aluminium, paper, cardboard, plastic, steel and aggregates. They found that in the vast majority of cases (83%), recycling was the most environmentally benign. They also estimated that the UK's recycling was saving 10-15 million tonnes of CO_2 equivalent (i.e. the amount of CO_2 that would produce the same warming effect as the other gas) which comprised about 2.5% of the total. This suggests that if we double the national recycling rate, which really means bringing every local authority up to the level of the current best performers, it would be the same as having a 5% cut in emissions.

If you don't presently recycle, this Labour is probably the easiest to fulfil as virtually every council offers a kerbside recycling service. If you do presently recycle, write to your council to ask them what plans they have to extend their schemes as some have now started to put recycling bins onto the streets for use by pedestrians. If your council doesn't currently collect plastic, urge them to do so.

Powering your House
(The Labour of The Ceryneian Hind)

Chronologically Hercules's First Labour had been the Nemean Lion and the Second the Lernaean Hydra, a fearsome beast we will meet later on, which he also slew. By the time it came to setting Hercules's Third Labour, Eurystheus had realised that just sending him off to kill things was no use, so he decided to spend more time thinking up a third task that would spell doom for the hero. This task did not involve killing a beast but he was instead charged with capturing the remaining Ceryneian Hind (the others had already been captured by Artemis, the goddess of the moon and the remaining free one was sacred to her).

The hind was so fast it could outrun an arrow. When Hercules awoke from sleeping, he could see the hind from the glint on its antlers and gave chase. Hercules chased the hind on foot for a full year before capturing it after it fell exhausted.

Although the Labour itself was fairly straightforward to complete, there is a subtlety that is appropriate for this part of the story. Eurystheus had hoped that Hercules would make Artemis angry if he succeeded and, indeed, Hercules did meet Artemis when he was taking the hind back to Eurystheus. Hercules had to beg the goddess for forgiveness, explaining that he had to catch the hind as part of his penance, but he promised to return it to freedom. Artemis forgave him, thus wrecking Eurystheus' plan. When

Hercules brought the hind to Eurystheus, he was told that it was to become part of the King's menagerie. Hercules knew that he had to return the hind to freedom, so he agreed to hand it over on the condition that Eurystheus himself came out and took it from him. The King came out, but the moment Hercules let the hind go it sprinted away, and Hercules departed saying that Eurystheus had not been quick enough. So by these means, a clever plan was derailed by an even cleverer hero.

When it comes to powering your house, the outcome you wish for is fairly simple: access to electricity. The subtlety of how it is generated, how much greenhouse gas is emitted in making it and how much you pay for it is more difficult to fathom. It is worth looking into, however, as this is an area where a big impact can be made very quickly.

In the UK, as in the rest of the world, generating electricity is the single biggest source of GHG.

The electricity market in the UK was liberalised thanks to an EU policy to open markets and since then there has been a flood of activity. Companies now vie with each other to win customers with seemingly ever more attractive deals and every company promises you savings. On the day of writing, the British Gas website homepage offered the opportunity to save an average of £134, E-ON offered a £177 saving, npower offered a £236 saving and Southern Electric offered 'even MORE money off your bill'. Each promise is followed by a little asterisk which leads you to some text that explains why is likely that *your* actual saving will be somewhat different, if there is any saving at all, but the small print keeps them out of trouble for misleading you into thinking you are going to get a great bargain.

Electricity prices vary by which part of the country you live in and which tariff you choose. Bizarrely, they can also vary by region and within tariffs. So, for example, I can buy electricity from npower for my house in London that is cheaper than my sister can buy for her house in Scotland. But there is a tariff where she would pay more than me for the first amount used (728 kWh to be precise, about 3 month's average usage) but then less for subsequent use!

Since liberalisation, it is estimated that 19 million people have switched their suppliers and price comparison websites have proliferated. Some people even made money out of this: entrepreneur Karen Darby founded Simply Switch and sold it to the *Daily Mail* for £22m in August 2006 only to see them close it 14 months later due to changed market conditions. Sometimes, timing is all.

Switching is easy. Figuring out if you have the best deal is pretty much impossible. This is due to the irreversibility of the decision. Once you have decided to switch supplier you will only ever know how much your supply costs *from that supplier*. You can't trial two companies simultaneously and see who works out the cheaper. Nor are the 'savings' easy to work out, as research by one of the leading switching sites uswitch.com found that 75% of consumers were confused by their energy bill.

I have been one of these switchers and, looking back over my direct debits over the past couple of years, I have found it impossible to work out whether I have saved any money by doing so. All I have noticed is that the bills have continued to rise regardless of who I am contracted with.

Our use of electricity has been a hotly debated topic over the past few years. Stories continually appear in the media about how we should be saving electricity, the evil that is the standby button and the global catastrophe being caused by plugged-in mobile phone chargers. This is another illustration of the confusion between energy and fuel. Electricity is an 'energy carrier': it moves energy from where it is produced, usually a power station, to where it is used, such as a television or stereo player. The fuel used to generate the electricity is critical to global emissions. Most of the UK's reductions in GHG emissions over recent years have come from switching some of our electricity generation from coal-fired power stations to gas-fired ones. Using gas to generate electricity generates about half as much emissions as using coal; using renewable sources like wind, tidal, wave, bio-mass and solar produces none. Whether you are an individual or a business, switching your electricity supply to 100% renewable sources is the quickest and easiest way you can make a substantial impact.

Unfortunately, there are a number of underlying confusions waiting to snare the unwary. The UK government has set a mandatory target of 15% of energy to come from renewable sources by 2020. This means that the proportion of electricity to come from renewables will have to rise from its current rather paltry 5.5% to around a third (because using renewable sources for transport and heat are more challenging).

Like a lot of government targets, the level seems fairly arbitrary and to ensure that it is achieved the government has developed a complicated system to monitor progress. Central to it is something called Renewable Obligation Certificates (or ROCs) which are presented to companies for every 1,000 units of electricity they generate from approved renewable sources. By comparing the number of ROCs to the total amount generated, you can see what percentage of the generator's output has come from renewables. Generators who have not met their target are obliged to buy certificates from those who have overachieved.

The Climate Change Levy was introduced in April 2001 to encourage businesses to improve their energy efficiency and switch to renewables and other more efficient sources of energy production such as Combined Heat and Power

(CHP). Levy Exemption Certificates (LECs) came about as a result of the Levy, and are allocated to generators for each thousand units of eligible power they produce. The renewable or qualifying CHP generator can enter commercial negotiations with suppliers to obtain value for the LECs they produce, which can be done separately from their power.

Renewable Energy Guarantees of Origin (REGOs) were created by an EU Directive in 2003. The certificates are issued by OFGEM to accredited generators for each kWh produced, as evidence that the electricity was generated from a renewable source as defined in the legislation. (OFGEM is the Office of the Gas and Electricity Markets. It is responsible for promoting competition in the gas and electricity markets and regulating the monopoly companies which run the gas and electricity networks.) REGOs are attached to the renewable energy generated and are used by suppliers in their Fuel Mix Disclosure.

Each of these schemes have their own regulations and processes that, surprise surprise, are slightly different from the others. It causes a bit of confusion, particularly for some of the smaller micro-generation projects that have been set up to try and get things moving a bit more quickly. A working party set up by the Department of Business, Enterprise and Regulatory Reform (BERR) concluded:

'Conceptually the simplest, but most radical, change would be for the micro-generation community (and/or their suppliers/agents) to be allowed to use REGOs as evidence for eligibility for ROCs (and possibly LECs) such that a holder of micro-generation REGOs would be able to request ROCs (and LECs) from Ofgem (1000 REGOs would qualify for 1 ROC and 1 LEC) without going through the accreditation process for ROCs/LECs or submitting additional meter data. The generator/agent could accumulate REGOs and exchange them for the equivalent amount of ROCs/LECs periodically. One potential difficulty is that the technologies eligible for REGOs/ROCs/LECs are not identical. However, REGOs currently specify the fuel source and a mechanism could be put in place to restrict eligibility to ROCs/LECs to appropriate technologies.'[38]

Thankfully, the more complicated solution wasn't outlined.

Even the Carbon Trust, an organisation funded by the government to accelerate the move to a low carbon economy, is confused and refuses to count moving to renewable energy tariffs as a reduction in carbon footprints![39] This strikes me as bizarre.

There are a number of different 'green tariffs' currently on the market from the major suppliers but they are not all the same. For some suppliers, it might mean that they supply you with renewable electricity, others invest in building wind turbines or in projects to offset household carbon emissions. Surprisingly, it could also mean that part or even all of your electricity may actually come from non-renewable sources. And when a supplier offers a 'dual fuel' deal, remember that 'gas' currently means fossil fuel and that means carbon emissions. British Gas had one of their adverts removed during 2008 because of the confusion they caused to consumers.[40]

The green tariffs are generally a bit higher than the regular tariff for each supplier, by about £1 or £2 per week. This is a truly amazing figure. Despite all the advantages of scale and written down capital enjoyed by the fossil fuel industry, renewable energy can be sold for the additional cost of a Sunday newspaper. What's more, given the variety of tariffs on offer, switching may actually be cheaper than your current supply. You can check your own situation at www.greenhelpline.com. Certainly when I switched from

Utilities Warehouse to renewable supply in 2009, my standing order stayed the same.

You can see the fuel mix of all of the generators at www.electricityinfo.org, an independent website providing environmental information on the UK electricity supply industry. There is only one UK supplier of 100% renewable electricity and they are called Good Energy. By switching your supply to them, you change the way your electricity is generated and strike at the heart of one of the biggest single sources of GHG emissions. What's more, readers of this book that switch will get £50 off their first bill. Simply phone 0845 456 1640 or visit the website at www.goodenergy.co.uk and when asked 'Where did you hear about us?' reply with 'Pioneer GEP0181'.

Why should you choose Good Energy over a green tariff from one of the major suppliers? Because it makes a bigger impact on the perception of investors and politicians and so speeds up the development of renewable generation (the major generators are currently arguing that they need to build more coal-fired power stations to 'keep the lights of Britain on'. This is nonsense as there are sufficient renewable resources even for Britain to become fully self-sufficient in terms of its power needs). In *Whoops*, as mentioned in Part One of this book, John Lanchester gives an excellent explanation of how investors look at numbers. If 100,000 customers of a major supplier with a total customer base of 4 million switch to a green tariff, then it 'only' accounts for 2.5% of the customer base and is sound evidence that people 'don't want to pay for going green'. If the same 100,000 sign up to Good Energy, it represents a 400% growth and instantly becomes recognised as the burgeoning enthusiasm for renewable electricity.

And remember, no use of fossil fuel = no GHG emissions.

Using Your House as a Generator and Keeping it Warm
(The Labour of The Apples of the Hesperides)

The story of Hercules fetching fruit from the divine garden of the goddess Hera might initially seem more suited to the next chapter on food, but I am going to use it here. Hercules didn't know where the garden was located and had to search for it in much the same way as we search for our ideal home. When he found it, it turned out to be a blissful place in the far western corner of the world tended by Atlas's daughters, the Hesperides. Atlas, of course, was a Titan who was destined to support the celestial world on his shoulders forever, in much the same way as we are burdened by mortgages on our homes. Hercules had been advised to ask Atlas to go and fetch the apples for him and temporarily take on his burden. When Atlas returned with the apples he was so pleased with his freedom he offered to take them to Eurystheus himself, but Hercules had been warned not to accept such an offer and had to trick him into taking back the globe so he could complete the challenge himself.

The fact that Hercules found the gardens in the west where the sun sets adds a magical dimension to the story. When the sun sets on a hard day's work and we set off for home, we usually do it with a sense of anticipation. Having a warm and welcoming home makes us happy but our homes are also important as a source of greenhouse gas emissions.

Heating our houses generates about the same amount of CO_2 emissions as our use of cars. What is quite different, however, is how much we spend on each of them. Whereas the average household spends £62 per week on motoring, the spend on fuel and power is about a quarter of that and represents only 4% of total household spend.[41] This means that fuel and power are likely to have a fairly low priority in most people's minds, which is a great pity as it is a valuable area for finding reductions in carbon emissions. Once again, we need to make a positive choice to engage.

The best way to engage with what is possible and appropriate for your home is to ask Atlas to go and get the apples for you: in other words, use a professional adviser. That is what I have done with the writing of this chapter and the material has been supplied by Jae Mather of the Carbon Free Group. Jae has worked in the environmental and sustainability sectors for over 10 years and holds a BA in Geography from Simon Fraser University and a Postgraduate Diploma in Environmental Decision Making from the Open University. He has worked with the UK and EU governments, as well as mainstream businesses, and has become a consultant in global environmental issues, eco-enterprise, environmental technology and sustainability. Although the Carbon Free Group focuses on the commercial sector, Jae has kindly provided the following material for individual householders. This chapter is necessarily a bit more technical than the others but it will help you to understand what any contractor you engage is talking about.

Energy and Heating

Energy and heating make up a large percentage of the total carbon emissions from a typical home with domestic energy consumption being responsible for 29% of the UK's total energy consumption (28% in 1990)[42]. As such, one of the main opportunities for carbon reduction in the home comes from reducing the need for fossil fuel based heating and electricity production.

> 'Britain has the oldest housing stock in the developed world with 8.5 million properties over 60 years old. It is therefore vital that we make improving the environmental performance of our housing stock a priority' (Energy Saving Trust)[43]

The Green Building Council's Low Carbon Existing Homes Report from October 2008[44] states that:

'The household sector represents 27 per cent of total CO_2 emissions in the UK, with most of the homes that will be standing in 2050 already built. Our homes have a significant role to play in ensuring the UK's climate change targets are met, particularly given the Climate Change Committee's recent recommendation of at least an 80% cut in CO_2 by 2050. High energy bills, rising fuel poverty and the need for energy security provide further justification to act.'

Heating and Transport Fuels

When it comes to heating, each fuel has a different impact in terms of the amount of GHG released. Fuels from wood and other plants are sometimes claimed to be carbon neutral as the CO_2 released when they are burnt is only that which was captured during the growth cycle. In reality, the harvesting, processing and transport of these fuels will inevitably involve the use of some fossil fuel so they should be considered as very low carbon fuel. When compared to the highest carbon fuels, coal or fuel oil, wood pellets allow for more than a 90% saving in GHG on a heat delivered basis. So if you do still have an open fire, switching to only burning logs would make an immediate impact in your carbon footprint. Of the fossil fuels, natural gas is the 'cleanest' although the GHG saving of natural gas over fuel oil is only 27%.

In regards to transport fuels, LPG fuel offers a GHG reduction of 11% when compared with Petrol (using kWh as a common comparison). Natural Gas offers a 19% reduction when compared with Petrol (per kWh) so converting your petrol engine to either of these does help. If you have a diesel fuelled vehicle it cannot be converted to LPG or Natural Gas, but it is possible to convert them to Bio-Diesel or Used Cooking Oil (UCO) which offer a great deal of possibilities for reducing carbon emissions. These fuels are somewhat controversial, however, as there is an issue of whether the fuel is offsetting food production and what its exact make-up is. Due to this, there are a wide range of fuel sources with differing carbon reductions.

Reducing the Need
Carbon reduction always starts by reducing the need for heating and energy as the starting step. Almost all buildings have opportunities for improving efficiency.

Insulation
Most existing homes can benefit from additional insulation. It is quite common for up to a third of the energy that is used to heat a house to leak out without giving any benefit at all. Every single property is different and even two properties of the same construction in the same street can have very different performance. Some local councils such as Haringey have started to make 'heat maps' available on their websites so that residents can see how well (or badly) they are doing in comparison to others round about them.[46] If your council doesn't provide such a service you could ask

them to think about doing so or you could simply have a look around the house.

The walls and the roof are the two leakiest parts of the house and between them are usually responsible for more than half of the heat loss. Usually this can be reduced significantly by applying insulation. This is easiest on houses that have cavity walls as they can be filled relatively inexpensively. Typically this can cost from £200 to 400, but that money can be recouped from reduced fuel bills with an average of 3 to 4 years (this is usually referred to as payback). Grants are often available from the Energy Saving Trust (EST), local authority or your current electricity or gas supplier. The EST website includes a search by postcode option that compiles all of the various offers together.[47] Many older houses have solid brick walls which would require either external or internal insulation to be fixed to the walls (you can tell how your walls are constructed by having a look at the pattern of the bricks. If they have little half bricks in the line they are usually solid walls. Cavity walls, which can be filled, have an overlapping pattern of whole bricks). If you choose to improve a house with the options available from the EST, it usually begins with the company sending a surveyor to your house to see what can be done and at what cost. The amount you contribute will depend on your actual circumstances and the process can be fully funded if you are getting certain benefit payments, but if you do have to pay it is still heavily subsidised. The actual work takes a couple of hours for loft insulation and about half a day for cavity wall insulation but be prepared for a bit of a time delay between making the enquiry to completing the work, generally around 1 to 2 months.

If you prefer to do the work yourself, several DIY retailers are working in cooperation with electricity providers to offer cheaper insulating material. There is also an organisation for those who want to eco-renovate their property to minimise their carbon impact. It has grown out of the work that George Marshall of the Climate Outreach and Information Network (COIN) did on his 1930s house in Oxford and now provides a support network to other 'ecovators'. You can find full details at www.theyellowhouse.org.uk/ and http://ecovation.org.uk/

There is a range of insulation materials available[48] in natural (low levels of embodied energy and generally high levels of moisture regulation) and manmade materials (with higher levels of embodied energy but generally cheaper), ultimately both main categories of materials are capable of offering very high levels of thermal performance with the main difference being different thickness levels and applications.

Glazing
Improving glazing is another key way of reducing heating demand and improving the comfort of a dwelling. Low quality windows also often lead to draughts and condensation build-up and improvements will thereby not only improve the heating efficiency but also the inhabitants' health as well.

Air Tightness
Draught proofing buildings is essential so that heat is not wasted. In the majority of cases, this is a fairly simple exercise that mainly involves upgrading the gaps around doors, vents and windows. Very high performance low carbon dwellings are extremely airtight. It is possible for very high performance housing to require little or no additional heating: the Passive House is a world leading system that specialises in such structures.[49] With very high levels of insulation and airtightness, they can generate enough heat

from TVs, kettles, white goods and the people who live in it, so that little or no further source is needed. Airtightness might sound suffocating, but the houses have a Mechanically Ventilated Heat Recovery (MVHR) system that completely changes all of the internal air five times an hour with fresh air while capturing over 90% of the heat in the outgoing air and returning it into the house.

Lighting
Building lighting is also responsible for a large amount of electricity consumption. In 1998, 17.5% of all electricity consumed in a house was for lighting but this is expected to drop to just 4% by 2050. This is due to Compact Fluorescent lighting (CFL) now being able to replace nearly all conventional incandescent lamps with light outputs of the same or greater levels and with the same colour rendering. It all comes down to purchasing quality CFLs when replacing incandescent lamps as low quality or inappropriate CFLs will not produce the desired lighting result. LED lights, which use even less electricity than CFLs, are developing at a very rapid pace and as such their quality, light output, colour and cost are all improving on a regular basis.

Standby
There is now a range of products that reduce the amount of electricity that is consumed when items are left on standby[50]. Bye Bye Standby home products are endorsed by the Energy Saving Trust and allow you to remotely cut power to electrical devices that are in standby mode. Research from Oxford University Consulting shows that Bye Bye Standby would save the average UK household £38 a year, which could represent as much as £940 million across the nation.

The Bye Bye Standby kit includes three smart sockets, which can control up to four electrical devices each, and one remote control. Simply plug Bye Bye Standby into a wall socket and then plug your appliance or extension lead into that. Each of the devices can be switched off separately or as a group when a room is left and pressing the remote button again instantly switches them back on. Research shows that people are not switching their appliances off at the mains. This may be because they forget or cannot be bothered to go round and switch everything off late at night or before they leave the house. Bye Bye Standby offers a convenient solution to switch off up to 12 devices at the touch of a button. Bye Bye Standby will also be helpful to older people or those with disabilities who cannot bend down to reach sockets.

Appliances
Energy efficient appliances offer a huge opportunity for saving electricity and reducing carbon. A+ or A++ rated white goods can actually offer paybacks when compared with old equipment due to the energy savings. As an example, an old fridge that consumes £60 per year of electricity can be replaced with a new A++ rated refrigerator that only consumes £20 per year thus offering a savings of £40 a year. With a purchase price of £200 there would be a 5 year payback period from replacing the old refrigerator[51] (see Table 1).

Savings assume replacing an average appliance purchased new in 1998 with an Energy Saving Recommended model of similar size and an electricity cost of 13.95p/kWh.

GHG savings will vary with each white good but typically high efficiency appliances will use 30 to 70% less than conventional equivalent equipment. Home Energy

Table 1

Appliance[52]	EU Energy rating	Saving a year (up to)	CO_2 saving a year (up to)
Fridge freezer	A+ or A++	£39	142 kg
Upright/ Chest Freezer	A+ or A++	£23	85 kg
Refrigerator	A+ or A++	£13	48 kg
Washing machine	A	£11	45 kg
Dishwasher	A	£23	90 kg
Integrated digital televisions		£7	24 kg

Monitoring has proven to be very effective when it comes to reducing electricity consumption. For some reason when people's energy usage is visible to them on a regular basis, there is a higher likelihood that energy saving behaviour will follow. Systems now are available for under £50.[53]

Feed In Tariffs
Feed In Tariffs (FIT's) are monies paid to householders who have fitted a renewable energy system to their home and are paid on the amount of electricity generated by the system. In 2011, there will also be a financial incentive for generating heat from renewable resources, some of which will be described later in the chapter. The Department for Energy and Climate Change (DECC) released a consultation document[54] entitled 'Consultation on Renewable Electricity Financial Incentives 2009', which outlines prospective Feed In Tariffs for a wide variety of

USING YOUR HOUSE AS A GENERATOR

renewable electricity generation systems. They are planned to be available for 25 years although the amount paid will fall by a small percentage each year. At the time of writing the actual tariffs had just been announced and the values offered appear to favour retro-fitting solar PV technology (described below) to existing houses. This will attract a price of 41.3p per kWh for the first year. Small scale wind turbines will attract 34.5p per kWh or 24.1p if the system is a little larger. Feed in Tariffs have been used in other countries for a number of years and have been very popular and successful in kick-starting the usage of small-scale local generation.

Of course, in order to benefit from a Feed in Tariff, you need to first of all install a system. It is a little surprising that the Government has opted to introduce incentives for electricity generation first as heating systems can generally reduce carbon emission by 50 to 60%. To reach the Government's 2050 target of an 80% reduction, technologies will be needed for electricity generation and perhaps they just thought the public would understand them more easily. It is important to stress that the costs associated with renewable energy generation decrease with the size of the systems.

The costs for district heating systems (where a large number of houses are provided heating from a single point) or large-scale electricity systems can easily drop by 10 to 40% when looked at from large-scale application. Wind power is a good example of economies of scale: micro-wind systems (defined as systems with a peak output of less than 5 kW, which is usually expressed as kWp. So if the system ran at its peak for one hour, it would generate 5kWh, but in reality systems generally run at anything between 30 and 65% of their peak) are largely ineffective in urban areas due to turbulence, obstruction or poor wind speeds. Medium-scale and large-scale turbines are often not acceptable in cities due to their larger size, but this is where economies of scale illustrate improved value.

The capital cost of installing micro-wind can cost from £5000 to £7000 per kWp installed, while medium-scale wind can cost from £1500 to £3000 per kWp and large-scale turbines can cost from £750-1000 per kWp. Due to these economies, large-scale wind turbines such as those installed offshore can offer renewable electricity generation with payback periods of 3.5 to 5 years, which is considerably less than a conventional power station. Large-scale systems are a much more cost-effective solution than individual house systems but the problem is finding the capital. Small-scale systems may be less efficient but the decision to install them is more easily taken. Outlined below are a number of individual house renewable solutions.

Photo Voltaic Thermal (PV-T)[55]
A PV-T collector is a combined assembly of:

- A PV module for the conversion of electrical energy
- A high efficiency flat plate solar collector for the conversion of thermal energy.

Photovoltaics (PV) are semi-conductors and suffer from a drawback: degradation in performance due to temperature. In the UK on a sunny day in the middle of summer, when you hope to be making the most of your PV array panel, temperatures can reach over 100°C. At this temperature the system will produce less than 10% of its maximum output, rendering it largely useless. By regulating panel temperature using a fluid cooling system, a balanced system can be produced trading off between

PV efficiency and thermal output. Using this principle it is possible to obtain a higher electrical yield: up to a 35% increase in electrical yield compared with the equivalent area of standard mono-crystalline PV and enough free heat to offset a low energy building's annual heating requirements. Due to these advantages, PV-T will in many cases offer much better value for money than conventional PV. In installations smaller than 2kWp, this may not be the case due to the fact that with PV-T there is a need for additional plumbing to carry the heating fluid but for larger systems it is proving to be a very attractive technical solution.

What can be achieved with a PV-T collector?
PV-T collectors allow for a 'total solar energy system', as both electric and thermal energy generation can be achieved. In combination with long term heat storage, this system can be used to take low energy buildings to, or near to, zero-carbon.

Examples of PV-T systems that could be retrofitted to dwellings: Please note that all calculations below are based on displacing natural gas central heating and electricity from the national grid. Dwellings that are heated with oil will offer shorter payback periods and larger amounts of carbon reduction. Also note that the values indicated for Feed In Tariffs are based upon the existing government proposals and that they are not yet finalised (in the proposed structure, there is an anticipated 7% reduction in payment per year). All prices for the systems below are based upon individual house purchases. Developers fitting in volume purchases would expect to reduce costs by 5 to 15%. All systems below also include appropriate hot water tanks for each system.

A 2 kWp PV-T system with a heat dissipater

Cost: £16,200 installed
Size: 16 Square Metres
Average annual Electricity production: 2034 kWh
Average annual Heat production: 6717 kWh

This system is able to produce approximately 70% of a renovated dwelling's annual heat and 40-50% of the electricity.

Average annual income from Feed in Tariffs
(at 36.5p per unit): £742
Average annual income if half of the electricity
is sold to grid (1017 kWh at 4.5 per kWh): £46
Average annual savings if half of the electricity
is used on site (1017 kW at 10p per kWh): £102
Average annual saving from the heat
produced (Natural Gas at 4.5p per kWh): £302.00

For a total value of £1192 per year giving a
7.4% tax free return on investment
Pay Back period: 13.6 years
Average Carbon Dioxide savings per year
if all of the electricity and heat is utilised
(electricity at 0.49 kg per kWh and gas at
0.204 KG per kW): 2346 kg per year

A 3 kWp PV-T system with a heat dissipater

Cost: £24,100 installed
Size: 24 Square Metres
Average annual Electricity production: 3235 kWh
Average annual Heat production: 11447 kWh

This system is able to produce approximately 100% of a

dwelling's annual heat (if inter seasonal heat storage is available) and 50-70% of the electricity.

Average annual income from Feed in Tariffs (at 36.5p per unit):	£1180
Average annual income if half of the electricity is sold to grid (1617 kWh at 4.5 per kWh):	£73
Average annual savings if half of the electricity is used on site (1617 kW at 10p per kWh):	£162
Average annual saving from the heat produced (Natural Gas at 4.5p per kWh):	£515

For a total value of £1930 per year giving an 8% tax free return on investment

Pay Back period:	12.5 years
Average Carbon Dioxide savings per year if all of the electricity and heat is utilised (electricity at 0.48 kg per kWh and gas at 0.204 KG per kW):	3957 kg per year

There is another version of PV-T that will be available that produces approximately 10% more electricity and approximately the same amount of heat. This system is thus a 1 for 1 producer of electricity versus heat whereas the regular PV-T is a 1 for 4 producer of electricity versus heat. This will offer an important alternative for buildings that do not need the heat due to very high levels of insulation.

Solar Hot Water[56]

What is Solar Thermal Water Heating?
Solar thermal hot water systems (SHW) are one of the simplest and most widely installed of the renewable energy systems in place today. They transfer the abundant energy

from the Sun into hot water in your hot water cylinder. SHW will even produce heat in the winter, but will require being topped up to the required temperature by a secondary heating source, such as your existing boiler or immersion heater. In this way, SHW complements your existing hot water system rather than replacing it outright.[57] For the average domestic dwelling in the UK, daily water heating consumes approximately 5 to 11 kWh of energy, but this varies considerably with household occupancy and lifestyle.

There are two main types of collector available, each with its own characteristics:

- Flat Plate/Flat Panel
  These offer high levels of durability, minimal maintenance (this only involves conventional pumps) and are less visually intrusive. They can be roof integrated which reduces planning issues.

- Evacuated Tube with either Direct Flow or Heat Pipes
  These offer higher levels of output in the winter than flat plate systems but with the added requirement of the need for greater levels of maintenance and associated fragility. They extend from the roof and thus are more visually intrusive.

Both types work by solar energy being absorbed by solar collectors, usually mounted on the roof of a building. When exposed to sunlight the solar collectors heat up to temperatures significantly above the ambient air temperature and once they are several degrees Celsius warmer than the water in the hot water cylinder, heat will be transferred to the cylinder.

Example cost of a typical flat plate system is:

A 4.5 Square Meter Flat Plate Solar Hot Water system

Cost:	£2800 installed
Size:	4.5 Square Metres
Average annual Heat production:	2055 kWh

This system is able to produce approximately 50% of the dwelling's annual hot water requirements of an average 4-person family house. Generation of heat will be incentivised by the Government from April 2011, but in the meantime Good Energy has introduced a scheme called HotROCs, which it pays for from the profits of selling natural gas to existing customers. That would mean this system would attract:

Average annual income from HotROCs (at 4.5p per unit):	£93
Average annual saving from the heat produced (Natural Gas at 4.5p per kWh):	£93

For a total value of £186 per year equivalent to 6.6% tax free return on investment

Pay Back period:	15 years
Average Carbon Dioxide savings per year (gas at 0.204 kg per kW):	419 kg per year

Solar Hot Water Tank[58]
Note that if you are installing one of the above systems it is likely that a new solar water tank would need to be retro-fitted as well. These special tanks are able to receive heat inputs from a range of supplies (e.g. Boiler and SHW) and are necessary for the system to work properly.

Typical prices for installed solar hot water tanks:

200 Litre	£1150
Suitable for 4-6 Square Meter SHW/PV-T	
400 Litre	£1300
Suitable for 8 Square Meter SHW or 16 Square Meter PV-T	
1000 Litre	£1900
Suitable for 24 Square Meter PV-T	

Air Source Heat Pump (ASHP)[59]

Heat pumps are a form of technology that works a bit like a refrigerator in reverse. They take heat available from either the ground or the air and 'concentrate' it for use in the home. Ground source heat pumps need to have their collectors buried. The amount of space needed is related to the amount of heat needed but generally they are not suitable to the small spaces available in UK urban gardens. A better option is to install an ASHP to provide space and water heating. Because they are taking heat from the air (either inside or outside) they are suitable for a wide range of applications, particularly retro-fit or where ground space is limited. The efficiency of air source heat pumps is slightly lower than ground source due to the constant fluctuation in air temperatures. However, air source heat pumps are designed to work at temperatures as low as -20° C so are still able to extract heat from the air at normal UK winter temperatures. Because they require no expensive ground collectors, air source heat pumps are considerably cheaper to install than ground source units.[60]

Typically, good quality ASHPs are able to operate with a COP of 3.3-3.4. COP is Coefficient of Performance which means for every kW of electricity used by the ASHP 3.3-3.4 kW of heat is produced.

In general terms units would need to range between 8 and 14 kW, the costs for installed systems would range from £4,700 to £5,500. These systems would be capable of heating a property of up to 150m sq as well as supplying hot water requirements. These systems can be used in conjunction with PV, PV-T or any other renewable electricity generation system so that the electricity required to power the units comes from renewable sources instead of the grid. It also means that your home could potentially be self-sufficient in heat and electricity, a consideration that may become increasingly important in the future.

When it comes to carbon reduction an 8 kW ASHP with a COP of 3.4 that is drawing electricity from the grid would be responsible for 3.84 KG (8 times 0.48 KG) of carbon emissions per hour of operation at full power to produce 27.2 kW of heat. If the same heat was produced from a gas boiler then there would be 5.55 KG (27.2 times 0.204 KG) of carbon emissions per hour of operation. Thus, this ASHP reduces heating emissions by 31% when compared with a gas boiler.

Geothermal Heat Storage[61]
One of the main challenges with renewable heat generation from solar systems is that the largest amount of heat is produced in the summer when it is not needed and the least heat is produced in the winter. Geothermal heat storage enables that heat to be stored in the ground so that the heat can be used when it is most needed. Solar Hot Water or PV-T systems can use geothermal storage to store the excess thermal energy generated in the summer to be used during winter. In effect geothermal storage systems allow for the creation of a thermal battery under the house.

Bio-Mass

Wood fuelled heating can come in a wide variety of forms, from log to briquette to pellet to chip fuelled systems. Systems can be used to heat a nearby space via radiation of heat or they can be central heating based boilers. In boiler form, they are typically more expensive than conventional gas based boilers but the fuel costs can be less expensive and the reduction in carbon emission is huge (up to 96% when compared with oil and just over 80% when compared with Natural Gas. Note that biomass is not carbon neutral as there is some carbon associated with the processing and transport of the fuel source). For a property currently off the gas grid and so using fuel oil or liquid natural gas, switching to wood pellet can offer payback periods of 8 to 10 years on average. And the fuel source is much more pleasant to handle than fuel oil.

There are an estimated 1.1 million new boilers sold each year in the UK so there are a great number of 'natural' opportunities to make a big impact on how much carbon UK houses are producing. There are a huge number of ways of decreasing carbon emissions in the home and improving one's own levels of self-sufficiency. New ideas, technologies and solutions are arriving on a regular basis and as the world comes to fully understand that business as usual is no longer an option, local solutions will once again become the norm when it comes to heating and powering our homes.

That's the end of Jae's section and I haven't had to trick him into going back to his own tasks. You might like to visit Jae's company website at www.carbonfreegroup.com if you would like more information on renewable technologies or to keep up with developments. The Carbon Free Group subscribes to the Declaration of Interdependence made by the David Suzuki Foundation for the United Nations Earth Summit in Rio de Janeiro. You will find a link to the Declaration from the Carbon Free Group's website home page.

Food
(The Labour of The Mares of Diomedes)

The mares of Diomedes were no ordinary horses, but rather had turned wild and hideous due to his strange habit of feeding them with the flesh of men. Although Hercules initially managed to steal away the horses, he had to fight to keep hold of them. Whilst Hercules was engaged in battle, the mares seized their chance and attacked and ate his companion Abderus. After winning the battle with customary ingenuity, Hercules stunned Diomedes with his club and laid his body in front of the mares, who proceeded to devour his still living flesh. Their appetites fully sated, Hercules had no trouble mastering them and delivering them to Eurystheus.

Food is obviously a powerful factor in our lives. It is vital for our health but can also affect our moods and plays a pivotal role in many of our ritual celebrations. What is 'good' and what is 'bad' for us generates almost as much confusion and debate as that on climate change itself.

Food production, delivery and consumption is an enormously important activity and a highly emotive issue. Modern humans have come an extremely long way from their ancestors when it comes to food. In the West in general and Britain in particular, we have 'outsourced' the production of food to specialists (agriculture employs only 1 in 100 people in the workforce) and has become increasingly detached from it. Surveys regularly appear in the media which demonstrate that quite high levels of people do not know where their food has come from (for example, the 2007 study from Linking Environment and Farming found that 22% of adults didn't know bacon and sausages originated from a farm, nor did 40% realise that yogurt was made with farm produce).[62] Conversely, when asked to imagine a farmer, people still tend to invoke the rural idyll by describing friendly farmers owning small-scale farms. The truth is very different. An estimated 80% of the global food supply is now in the hands of just five corporations.

Our relationship with food in many ways reflects our confusion about whether we are the master of nature or simply merely part of it. Although 2008 and 2009 have seen a comparative boom in the 'grow your own vegetables' movement, the majority seem to think that the best food betrays no connection with any living organism: an immaculate conception that lands miraculously in the supermarket chiller cabinet, gift-wrapped in shiny, clean packaging.

For most of the 1990s, there was a boom in cookery programmes on TV and glossy food magazines whilst home cooking declined, although the effects of the recession seems to have partially reversed the trend.[63] This curious relationship with food also points us to an odd place to look for the first step in reducing our impact: the dustbin. The fact is that we throw away perfectly good food at the rather alarming rate of about 100kg per person per year. That's about twenty full shopping bags worth of food a year. Whether it's a couple of chicken breasts which are past their use by date, or the last few slices of bread which have gone stale, or a chunk of cheese which has gone mouldy, collectively we are chucking away about a third of all the food we buy.[64] Some of that waste is inevitable, such as peelings or bones, and some is unavoidable particularly if you have a household in which plans change frequently. But clearly a lot of it could be avoided. The government backed body, WRAP, the Waste Resources Action Programme helps people reduce waste and recycle more and they estimate that if everyone avoided this unnecessary wastage it could have the same impact as a 20% reduction in the car fleet.

The second place to look for an impact is your waistline. Almost a quarter of adults are classified as obese. Obesity is now measured as a combination of your Body Mass Index (a measure calculated from your weight and height) and waist measurement. This measurement avoids the previous anomaly of very fit people being classed as obese due to having a high weight because of increased amounts of muscle. I'm not going to cover the details of the measurement here, because I think you probably already know if you are carrying too much weight. If you are, then the first step towards reducing it is to decide it is a problem you want to take action on. If you have already made a positive choice to take action on the climate change issue, losing weight by eating less is simply an extension of that process so perhaps that will spur you on.

When it comes to actual food production, an inevitable side-effect of the industrialisation of the food system has been that the industry is dependent upon fossil fuels. Producing, processing and transporting our food takes an enormous amount of energy: some have estimated that it takes 6 to 9 times more energy than the amount contained in the calories in the food on our plate. Just how much GHG is generated by the food system is difficult to estimate. The complexity of the system and the variety of people's diets make it almost impossible to calculate with accuracy. But it is almost certainly the single biggest source of emissions on the planet. Which, given the importance of food in keeping us alive, is a bit troublesome. This is one area where, in order to be a Humankind Superhero you need to be prepared for some radical action.

Let's start with the ground. In order to grow plants, the ground needs to have nutrients and water. One of the most important nutrients is nitrogen, the element that makes up about 80% of the air that we breathe. Put more nitrogen in to the soil and the plants grow more abundantly. Industrialised farming uses fertilizer made from oil, itself an energy-intensive process plus of course it needs to be transported from the factory to the farm. When these fertilizers break down they release the same nitrous oxides that we discussed back in the transport section. The production and breakdown of inorganic fertilizer is the single biggest source of GHG in agriculture.

Organic farming methodology uses clover, a plant, to extract nitrogen from the air and fix it in the soil, which is topped up with animal manure. It appears that some nitrous oxide is still released when the fields are ploughed, but nowhere near the levels of that produced by the fertilizers. This means that buying organic food ranks highly in the list of possible actions that will reduce your carbon footprint. Remarkably, the organic food industry has been slow to communicate this dimension, having previously simply argued that organic was better for the environment. They also argued that organic food was better for your health, but this is a highly contentious claim and there was a heated debate in 2009 when the UK Food Standards Agency said the nutritional value of organic food was no different to non-organic. In fact, the Food Standards Agency, when asked whether organic food is 'safer' than conventional food answered: 'Both organic and conventional foods have to meet the same legal food safety requirements'. This is rather a good example of the blinkered thinking that people easily slip into, as it confuses the safety of merely eating the food (such as it will not poison you) with the safety of the production process.

A diet of organically farmed grains, nuts and pulses is the most effective way to reduce GHG and these foods have an environmental impact of around a third or less in comparison to an omnivorous diet. This means that, if we all only ate plants, in theory three times as many people could be fed and the environmental impact would remain unchanged. A diet that is entirely based on plants is called vegan but very few people follow it (about 0.5% of the UK population).[65] They are a living demonstration of the possible and I respect them greatly for it.

So can you still be a Superhero if you remain an omnivore? There are very serious concerns about the impact of animal farming, particularly when it is performed with the intensive methods used in the developed world. Some of the issues, however, are straightforward to address. Pigs, for example, excrete four times as much as humans do in terms of quantity, but farms rarely treat the sewage. So a pig farm of 5,000 hogs could be dumping as much untreated sewage as a medium-sized town. Treating the sewage and using the methane to generate power would be an impactful change. Others problems are much more complex and challenge our assumptions about what it means to be a carnivore. One thing is for certain though: all animals are equal, but some are more equal than others when it comes to considering their environmental impact.[66]

Wild fish are probably the least damaging in terms of GHG emissions, as it is 'only' the fuel used by the fishing boats that catch them and their subsequent transportation to market that are carbon producing activities. Unfortunately, the seas are not nearly as full of fish as they once were. An advert for *The Times*, which ran during the autumn of 2009, claimed that the oceans will be empty of fish in 41 years. Perhaps more accurately they should have said that humans will have taken them all.

Farming fish is controversial, not least because they need a lot of animal protein (usually from other fish) to feed them. There are vegetarian fish such as carp and tilapia so the farming of them should be more benign. This is not the case, however. A recent study by Stanford University in the US found that farmers of these vegetarian fish had been adding fishmeal to boost yields and that even though it was only a small amount in terms of the total feed, the scale of the operation meant that, on a global scale, farmed 'vegetarian' fish are eating more fishmeal than the farmed carnivorous salmon and shrimp.[67] The main argument for including fish in your diet is for the consumption of essential fatty acids, but these can be gained from vegetable sources like flax seed, so when all is taken into account it is probably best to avoid fish altogether.

When it comes to land based animals, the first thing to remember is that the benefits of organic production remains just as valid here as it does for plant crops. An animal that naturally grazes or forages will have about half the impact of one raised using conventional means. Unfortunately, less than 1% of UK meat production is currently organic and the majority of the fertilizer used in the UK is for grass land. The second point is that there is a big difference between what are called ruminants, such as cattle and sheep, and other farm animals. When ruminants digest grass or other fibrous material, they produce methane, a greenhouse gas that is about 25 times more powerful than CO_2. Sheep are the worst here: they are effectively little methane bombs as they produce more than 3 and-a-half times GHG as poultry per kilogram of meat. And remember, not all sheep are raised for meat. Many are farmed for their wool.

While mutton was once a popular dish, it is virtually impossible to find now but presumably ends up in pet food. Cattle are the next biggest culprit, producing a whopping 16kg of CO_2 equivalent for every kilogram of meat. Cattle, of course, do not only produce beef, but are also used for dairy products, but whether they are being grown for beef or milk, they still burp constantly. So much so, a vegetarian diet that includes dairy produce is unlikely to be any less impactful in a GHG sense than a diet of vegetables supplemented with chicken and pork.

The total GHG impact of your diet will change according to the exact proportions of vegetable, dairy and white meat you include as well as how they are produced. If you like you can have hours of argumentative fun over who has the 'best' diet but I hope you catch the general idea.

Globally, the demand for beef is rapidly expanding and a lot of deforestation occurs as a result of producing soya to feed the cattle. This delivers a double whammy to efforts to cut down carbon emissions: we are causing deforestation in order to feed the cows that are themselves a major generator of the problem. It can only be considered analogous to putting out a fire by pouring petrol onto it. And remember, red meat occurs in processed foods like beef burgers. It might seem strange, but choosing a KFC meal over a McDonalds or Burger King is probably a better carbon choice.

If you cannot do without your red meat, at the very least trying to cut out two or three days' worth per week and only buying organic would be an impactful step. 'But isn't it really expensive to buy organic meat?', I hear you ask. Yes it is and this is where you need to start thinking about your thinking. Marketing messages like to focus upon a simple idea and repeat it endlessly. That is why you see so many advertising headlines with 'save' or 'free' in them. They want you to only focus on a single attribute of that product, so that you can justify your purchase decisions in your own mind, usually on the basis of price or luxury. What I am suggesting here is a little different. By stepping up your thinking, you should be able to calculate the effect of multiple decisions at the same time.

Our family is fairly typical in that we do a weekly shop at Tesco. We cook most of our meals from scratch which is somewhat less common than the average, but until a year ago I shopped pretty much like most people, looking for offers and buying stuff without too much consideration of the origin. Our weekly shopping bill for a family of four is about £165. Since taking on the project of writing this book, I have become more aware of what I am buying. At the time of writing, Tesco offered a range of choices of chicken and I wanted to buy one for our Sunday roast: there was a nice and plump free-range organic one at £11.82 or, at the other end of the price range, there was an offer of three chickens for £10. However, because I was doing a weekly shop and we had already planned two meals based upon vegetables, I decided to go with the organic bird.* The total bill for the shop was £163, pretty much in line with the average for our household.

Incidentally, if you want to start bringing more-vegetable based meals into your diet but don't know where to start, a book by the Australian *Women's Weekly* called *Almost Vegetarian* is a good guide. It gives a host of

---

* And quite delicious it was too! I took it out of the fridge an hour before cooking and rubbed oil and lemon pepper into the skin. Roasted, served with roast potatoes and curly kale. Even my sceptical teenage son was impressed with the chicken.

vegetarian recipes with a variation to include meat or fish, so you can try them out then repeat the ones you like in their original vegetarian style.

The next thing to consider is where the food is produced. It would be nice to be able to tell you to only buy what can be grown locally, but that would diminish us and, in some parts of the country, restrict our food choice to a fairly boring selection. Food was always taken by conquerors to their new lands. The Romans brought grapes and many other foods to Britain. Food was also amongst the earliest of items traded along the spice routes, with teas from China adding an exotic dimension to European society. Once you have grown use to a varied diet, restricting too narrowly seems like deprivation.

The location of production also needs to be considered but it is not always obvious what the best choice is from the standpoint of GHG emissions. Tomatoes grown in Scotland in fossil fuel heated greenhouses (as they were until the late 1970s) are no match for those raised in sunny Spain or Italy, even when the transportation is taken into account.

As a general principle, buying local produce during the season in which it grows best is the most impactful strategy, but the reality of today's market makes it difficult to implement. This is particularly true of fruit and vegetables, which may be why there has been such a strong upsurge in home growing and waiting lists for allotment spaces. If you do decide to engage in home growing, obviously you need to ensure you use organic fertilizer for your plants, which probably means shunning most of the products in the local garden centre.

In the case of meat it is much easier. An animal will produce the same amount of GHG whether it is in Scotland, England, Ireland or Argentina. Transportation is the added carbon cost. When buying meat, choose local.

So where should you shop? This is a controversial issue. The traditional green view is that we should support local independent shops and find a village high street where we can walk from specialist to specialist gathering our prized purchases. The urban world has concentrated these specialists into the departments of the supermarkets. These enormously popular spaces are, I think, designed to appeal to our hunter-gatherer instincts. A huge amount of time, effort and money goes into designing these superstores and our shopping experiences in them, so we become more and more willing to part with our cash. But I'm OK with that. Trade, remember, is where both parties come away feeling good about the transaction. For me, the range, convenience and prices supermarkets can offer makes it a valid trade. And I'm not alone. So how can we get the supermarkets that we enjoy using so much to contribute? They already are. Marks & Spencer have been working for some time on their 'Plan A' to reduce the environmental impact of their operations. Tesco has said they want to bring the green movement into the mainstream and are working on carbon labelling for their products and offer 'Green points' as part of their loyalty scheme to those who bring their own bags or purchase specific products. In truth, it is virtually impossible with current information to say which supermarket is 'best'. Many supermarkets now offer internet purchases and home delivery. This is a good thing in so far as it at least eliminates the transportation element of people driving to the store to select their goods, replacing this with a delivery van making multiple drops. However, it is still only a minority who do their shopping this way, comprising less than 3% of total groceries purchased.[68] Maybe hunter-gathering on the screen just isn't as exciting as being there in the flesh? There is, however, a half-way house that no one seems yet to have investigated. Travelling to the supermarket of our choice will almost always involve a journey of only a few miles, which is certainly within cycling distance. Cycling to the supermarket doesn't work out particularly well in reality, as it is very difficult to get the booty home. What I would like to see is a supermarket that would let me 'pick and drop'. So, I would cycle to my local store, enjoy the full shopping experience and then they deliver the goods to my house. I have written to ASDA, Tesco and Sainsbury's to suggest they may like to trial such a scheme and if you want to support the idea, you will find their addresses at the back of the book.

The Shopping Habit
(The Labour of The Cretan Bull)

The seventh Labour of Hercules was to capture the Cretan Bull and is appropriate for our seventh Labour. Although the origin of the bull varies from version to version of the myth, the underlying story is that Hercules travelled to Crete where the king offered him every assistance in his power, but Hercules declined, preferring to capture the bull single-handed. Triumphant, he took the bull back and presented it to Eurystheus who in turn gifted it to Hera. However, as the gift glorified Hercules, she refused it and it was set free.

Since the dawn of thought, philosophers and mystics have told us that wealth and fame are antithetical to life's true meaning. But, despite these wise men offering us every assistance, we find this difficult to believe and want to find out for ourselves. And what better place to find out than in the shops? John Lewis, the retailer, probably hit it on the head with their advertising campaign that said simply: 'We believe nothing should get in the way of that "Ooh, look what I've just bought" feeling.'

There is a long human history of adornment. There is no simple way to know for certain what the intent behind cave painting was, but they can be seen as the first evidence of our need to beautify our surroundings. The history of art and design is a history of our attempts to fulfil that need. Few but the most avid of ascetics would disagree that the quality and setting in which we live our lives affects the quality of the lives we live. Given the choice between a hillside shanty town and fine suburban house there is little doubt about which most of us would choose.

The modern 'consumer society' probably has its roots in the 1930s. This is a decade that mostly gets portrayed as a gloomy time: unemployment, poverty, the rise of fascism and a looming war. That certainly was one side of it. But even though unemployment reached 25%, this still meant that the majority had jobs. And for these people, things were rather better: the national grid was completed to bring power and new possibilities to a mass market; holiday pay was introduced causing a boom in seaside towns; millions of new houses were built, most of them for private ownership: new labour-saving gadgets like washing machines became available which relieved the drudgery of household tasks and 'hire-purchase' became prevalent, allowing the acquisition of goods with an immediacy never before imagined. It was also the decade in which marketing as we know it was invented. In a book published in 1932 called *Consumer Engineering: A New Technique for Prosperity*, Earnest Calkins wrote: 'Goods fall into two classes: those which we use, such as motor cars and safety razors, and those which we use up, such as toothpaste or soda biscuits. Consumer engineering must see to it that we use up the kind of goods that we now merely use.'

A combination of 'consumer engineering' coupled with retailers' obsession with separating customers from their cash has led to the development of an economy designed to constantly prime the pleasure centres of the brain, to keep us lusting after things we don't need. Retailers know that even though we probably won't buy the most expensive items on display, just looking at their beautiful displays of top-end items make us more likely to buy something else, since the coveted items activate the brain's pleasure centres.

But it's not enough to just excite our reward centres; retailers must also inhibit the part of the brain responsible for making us cautious. We don't want to spend excessively, so all but the most exclusive of retail stores repeatedly assure us that low prices are 'guaranteed' in order to stop us worrying so much about the price tag. In fact, researchers have found that even when a store puts a promotional sticker next to the price tag, something like 'bargain buy!' or 'special offer!', but doesn't actually reduce the price, sales of the item will still dramatically increase.[69]

People shop because they have a problem they need to solve (I need food for the family), or because they want to feel good, or both. Shopping's apparent ability to make us feel good has even caused it to become known as 'retail therapy'. This becomes a never-ending task, however, because we soon become accustomed to our newly acquired goods and the pleasure from them tends to fade quickly. Yesterday's luxuries have a habit of becoming today's necessities and, thereafter, tomorrow's cast offs. Or, as a poster in Selfridges once said: 'You want it, you buy it, you forget it.'

The purpose of this book is not to moralise about whether shopping is right or wrong. There is no singular formula to determine what life's pleasures are, nor should there be one. The relationship between shopping and the rest of one's life is a complex one. The money which we use to facilitate our civilisation only has value when it is moving, and buying things is the easiest way to make it move (of course, the real complications have emerged in the past few years after people were very much encouraged to buy things with money they did not have, but that is another story). The objective here is to give you a strategy for minimising the carbon impact of your shopping while living within the mainstream of society. This is actually the area where your actions can have most impact, as making, moving and selling us stuff is the single biggest use of energy in the UK.[70]

The concept of shopping sustainably is not a new one: the Fairtrade brand has been in operation for 15 years. However, their rate of uptake has not been as fast as, say, the growth of the internet, and 'sustainable' products are presently only believed to account for 4% of UK consumer spend.[71]

Global warming has also been discussed for a number of years and people are now reporting high levels of anxiety about the future. So what causes this apparent contradiction between being worried but taking no action? Partly it is down to price. Remember that buying something is a trade off between the pleasure of making the purchase and the pain of parting with the money needed to acquire it. A study by PWC[71], a major consultancy, found that nearly half of their respondents were either unwilling or unable to pay the premium associated with more sustainable goods. When asked how much they would be prepared to pay extra for sustainability, people tend to think about 20% as being a fair premium. But in a price comparison shop of 75 items at the top six UK grocers, environmentally and ethically friendly products were on average 45% more expensive than their standard counterparts.

There is also a lot of confusion and not a little distrust surrounding the whole area. The contradictory and often overwhelmingly complex information available about the implications of selecting one product over another leaves consumers confused. In the face of confusion, the natural tendency is to stick with what you know or simply do nothing. People want to make sustainable choices, but are hampered by unclear messages.

The government response to this is to provide more

information and, through the Carbon Trust, is sponsoring the idea of carbon labelling. You might have noticed some of the products in your shopping basket have started sporting a small black footprint logo. Tesco's fresh orange juice and Kingsmill bread provide examples of such labelling. These carbon labels are intended to show customers the 'carbon content' of an individual product. An item's carbon content is the total amount of carbon dioxide emitted from every stage of its production and distribution, from source to store. This is also known as 'embedded carbon' or a carbon footprint.

Having three different terms for the same thing doesn't usually help to clear confusion. Early reactions from companies running pilot trials, such as Boots, show that carbon labels are yet to impress customers, who often fail to understand the information on them (according to a survey of Boots' customers). Another survey found that just 28% of Boots' customers knew that a product's carbon footprint was related to climate change. And 44% confused it with Fairtrade.[72]

Another problem was that while few products carry the labels, even clued-up consumers could not compare like with like, or were able to judge whether 200 grams of carbon was high or not. The Carbon Trust remains upbeat about the potential for the scheme and the idea is that if a company does not manage to reduce the carbon content of a product within two years they will lose the right to use the label. However, that might be a better incentive for the manufacturer than the customer. Defending the scheme, the Carbon Trust likened the measure to calories and claimed that twenty years ago if you asked somebody what number of calories was high and what was low, they would not have known. Unfortunately, knowledge of calorie numbers has not prevented an alarming rise in the incidence of obesity in the country!

A modern supermarket reputedly stocks 40,000 product lines. Not all of them have sustainable alternatives available. While in a grocery store, you may be able to find sustainable alternatives for most of the basics, but there are many categories of food in which this is much more difficult. For clothing and other non-food items, there quite might be no alternative and the choice is simply amongst a range of standard items. Trying to work out what is the 'best buy' is a minefield and way beyond the scope of what can be covered in a single chapter. However, there is a plethora of green and ethical consumer guides available. Julia Halies' *The New Green Consumer Guide* is a good one, as is Leo Hickman's *A Good Life: The Guide to Living Ethically*. Reading them does not amount to taking up an 'alternative lifestyle'. Nor is it about 'saving the planet'. It is about adopting a smarter way of doing things and finding a way of living that makes more sense. Eventually, everyone will have to do it, so this is your chance to be ahead on the fashion curve. What follows is really just a summary to give a broader understanding. We covered food and cars in previous chapters, so the following really concerns those areas that economists like to call 'discretionary spend', which refers to the purchase of products that we only need *some* of but either prefer or somehow end up buying more of than we need.

Services are generally less carbon intensive than physical items. The bigger the proportion of the price that is related to labour, the lower the carbon impact will usually be. So, if you feel the need for a little 'retail therapy' head down to the spa rather than the shopping centre. There is a shopping centre near where I live that has welcoming posters on its doors that illustrate a pair of shoes with the slogan: 'Because you only have 19 pairs of red shoes'. Rather than buying another pair of shoes, why not have a pedicure? The possibilities here are nearly endless. Instead of buying a thing, buy a person's time. What could be nicer than having someone do things for you, rather than worrying if your friends will approve of how you look in your new outfit? There is even research that suggests that buying experiences rather than things leads to more happiness.[73] The opportunity to share experiences with other people and talk about them later is better for you and probably less carbon intensive than that 20th pair of red shoes.

When you do buy something, think second-hand first. Searching charity shops and website listings can be just as much fun and appeals to the hunter-gatherer instincts just as well as going to the out-of-town shopping centre. Buying second-hand doesn't reduce the amount of carbon that has been produced when making the product but it does postpone the release of more carbon incurred in buying a replacement. The amount of carbon released by transporting stuff to a charity shop or boot sale is much less than making (and, probably, transporting from China) a new one. It will probably save you money too, but you need to be careful what you do with the surplus cash. Don't be tempted to spend it on extra travel as that would simply destroy your good work. Some more trips to that spa might be nice, though. When you do buy a new thing, buy the best you can afford. The production of luxury clothes, shoes or gadgets does not generate much more carbon than a cheap version. It is the quantity of stuff that you buy that drives the carbon output, not the amount you spend. Choose wisely and you also get the kudos of the premium product.

We use the term 'taking the bull by the horns' when we need to grapple with particularly thorny problems, which brings us back to the Cretan Bull. Figuring out our personal relationship with the things we buy is a big step towards personal empowerment and freedom. Are we using the things we buy to try and glorify ourselves? Are we seeking the approval of others through our brand choices? Are we trying to emulate the lifestyle of the rich or famous? Being aware of ourselves and standing by our own decisions is worthy of true Superhero status.

Plant Trees
(The Labour of The Erymanthian Boar)

The Erymanthian boar was named after the mountain it lived on. It was a fierce, enormous beast and every day the boar would come crashing down from his lair on the mountain, attacking men and animals all over the countryside, gouging them with its tusks and destroying everything in its path. To take such a beast alive was going to be difficult, but finding it was easy. Hercules could hear the beast snorting and stomping as it rooted around for something to eat. He chased the boar round and around the mountain, shouting as loud as he could, until the boar hid in a thicket. Hercules poked his spear into the thicket and drove the exhausted animal into a deep patch of snow. Then he trapped the boar in a net and carried it to Eurystheus, who was once again amazed at the strength of the hero.

Using the environment to help us achieve our goals is a pretty ordinary occurrence for humans. Forests and woodlands are important parts of the landscape, both in the UK and elsewhere. As well as being beautiful and fun things to climb when you're a kid, trees are hugely important for the planet. During the day, they soak up carbon dioxide from the air, which they use to help them grow. The carbon turns into wood and the oxygen gets released back into the atmosphere. They really are the lungs of the planet.

They are also very useful to humans whether it is by

means of providing fuel for cooking or heating, or building materials, or being transformed into paper or other products.

Trees cover about 30% of the Earth's total land area, but the forests are unevenly distributed around the world, with just 10 countries possessing two thirds of the total, whilst 64 countries have less than 10% of their land area as forest cover. Just over a third of the world's forests are truly wild places with no clearly visible indications of human activity. Only 4% are forest plantations, growing trees to order mostly for the paper industry. The remainder of the world's forests are a somewhat haphazard alliance between people and plants, supporting the livelihood of an estimated 1.6 billion people. Some of them mange to do this sustainably, but an awful lot do not.

Generally speaking, the richest biodiversity, measured in terms of the variety of plants, birds and other species in any given place, is to be found in the wild forests that are remote from humans.

Broadly speaking, the world's forests grow in two great lateral bands, one stretching across northern latitudes and incorporating the forests of North America, Scandinavia and Russia. This band is of the type of forest known as 'temperate and boreal'. These contain the type of trees that are most familiar to people in Britain, with a mix of broadleaf trees such as oak, ash, sycamore and chestnut along with evergreen and needle leaf varieties such as pine, spruce and larch. The other major band runs across latitudes in the southern hemisphere and incorporates the forests of South America, Central Africa and Asia. These are tropical forests and are not all rainforests, as some of them are at higher altitudes or by the coast where they form mangrove forests. Mangroves are particularly important. They are tidal forests and have important functions as natural sea defences, nurseries for fisheries, and habitats for lots of other species.

The probability of sea-level rises and extreme weather events caused by climate change raises the importance of mangroves as a buffer protecting coastlines in the tropics and subtropics. Despite this, mangroves worldwide have been subjected to an appalling rate of destruction resulting from over-harvesting for timber and fuel wood, clearing for shrimp farms, agriculture, coastal development and tourism.[74] Mangroves have been destroyed much faster than any other forest type.

Forest exploitation, just like fossil fuel exploitation, occurs in line with the same economic system that pays no price for the cost of environmental destruction. Indeed, destroying forests for timber is big business, with the global value of wood imports worth $160 billion in 2006[75] and the rate of cutting them down outstrips the rate of replanting by about 7m hectares a year[76] (which is the space occupied by around 85 billion trees).

Although forests have lots of different possible uses, policymakers, particularly in the developing world, often do not consider forests to have a value other than timber, and defend their exploitation on the basis that the developed world destroyed their forests years ago as part of the development process. Besides timber, forests can also produce other direct use products such as latex, cork, fruit, nuts, spices, natural oils and resins, and medicines. Many of the medicines we use today have come from forest products and nobody knows what else may be discovered.

Forests can also be used for recreation and even spiritual respite. As these uses are related to the existence of a range of tree, plant, animal and other species, forests have an important role in providing habitat for the preservation of these species, particularly in tropical areas.

In fact, tropical rainforests contain a phenomenal range of species, more than twice as many as any other forest type and many more of them are unique to their own forest. Forests also have important benefits for the countries in which they are located in terms of recycling nutrients in the soil and providing watershed protection. Forested watersheds act like a sponge that slowly lets out the water so providing a more constant water flow into the rivers and so reducing floods. Cutting down the forests also leads to the soil being washed away, taking its nutrients with it and leading to build ups of mud in water reservoirs and rivers. What's more, forests have a big impact on climate both locally and on a wider scale. Local rainfall can be reduced once a forest has been cut down because the sponge dries out and the trees are no longer giving out water vapour.

Lastly, of course, once you have cut down a tree and turned it into timber, it is no longer breathing and removing carbon dioxide from the atmosphere. And that is a big contributor to the global greenhouse gas problem. Not only have we been putting pressure on the atmospheric system by pumping out extraneous gases from industrial, transport and farming activities, we have been cutting down the lungs of the planet at an alarming rate. So much so that around a fifth of the GHG problem is due to deforestation.

Poverty is usually cited as *the* underlying cause of tropical deforestation, but it is more complicated than that. Poverty does drive people to migrate to forest frontiers, where they engage in slash-and-burn forest clearing. They do this because it leaves a layer of ash that acts as a fertilizer for subsistence farming. This is seldom the whole story, however, and subsistence activities are estimated to be responsible for less than half of the destruction of forests.[77]

State policies to encourage economic development,

such as road and railway expansion projects, have caused significant, even if unintentional, deforestation in the Amazon and Central America. Agricultural subsidies and tax breaks, as well as timber concessions, have encouraged forest clearing as well. Global economic factors also pile on the pressure: countries burdened by foreign debt are tempted by the growing world market for rainforest timber and pulpwood. Faced with an opportunity to make substantial sums of money, few companies or governments turn it down. It needs a high order of positive choice to take the low carbon route. The actual rate of deforestation is difficult to measure because lots of countries use different methods of defining what a 'forest' is and the world certainly cannot agree what a 'degraded forest' is.[78] Even when you see the estimates, researchers are talking about areas so huge they are difficult to comprehend: maybe an area the size of England every year for the past 15 years. About a third of the world's total rainforests have gone in the past 50 years.

There is less uncertainty about the main areas of deforestation, however, with large areas of the Amazon being lost to industrial-scale cattle ranching and soybean production. In Indonesia, the conversion of tropical forest to commercial palm tree plantations to produce bio-fuels for export is a major cause of deforestation on the islands of Borneo (which is mostly administered by Indonesia) and Sumatra. Brazil and Indonesia may have been the giants of deforestation between 1990 and 2005, but Sudan, Myanmar, and the Democratic Republic of Congo have since also deforested large areas[79].

There is a growing recognition of the importance of forests in general and the tropical rainforests in particular. There is also a growing recognition of the damage caused to the global environment by the false economic assumptions regarding environmental costs.

Europe has introduced a carbon trading scheme that seeks to introduce a cost to polluting. Carbon markets, as they are called, are highly controversial and still in their infancy. The idea is to bring down emissions by making it more expensive to make them. If you cannot cut your emissions, you need to buy more 'carbon credits'. In theory, if the price of carbon credits is high enough, you will choose to make an investment in a low-carbon project instead. Although, the market is already substantial in Europe, the price of carbon credits has not yet risen high enough to instigate the scale of low-carbon investment that policymakers had hoped for. There is quite a debate between environmentalists and businesses, which is centred upon who will make the money from these carbon credits and how quickly they will function effectively. Forests are an important part of this debate, as you can currently use their carbon-capturing abilities to create carbon credits that can be sold on the European market. However, rainforest destruction continues at a frightening pace.

Alongside the efforts to create a market, there are now discussions taking place at the United Nations to place a value on existing forests. The idea is to make the carbon dioxide stored in the forests into a commodity that can be bought and sold on the global market. For the first time, existing tropical forests would be worth money for simply being there. That could create an economic incentive for governments and companies to protect their tropical forests. This is potentially very good news but currently the actual details of how it would work and when it might start are not available. Given that it is a political process and has the deadening weight of an industry with vested interest to fight against, it is worth writing to both your MP and MEP to encourage them to campaign for its development as a priority. If you don't know who they are, you can use the website www.writetothem.com. This is a non-charging website that will identify your representative and allow you to send an electronic letter directly to them. The scheme you are asking them to support is called Reduced Emissions from Deforestation and Degradation in developing countries, or REDD. Additionally, you might like to support the Prince of Wales' Rainforest SOS initiative. This seeks to create a temporary global institution and framework to act as a bridge towards a long-term UN solution, and also to establish a funding mechanism that draws on the combined strengths of the public and private sectors to channel money towards stopping rainforest destruction immediately. It is an ambitious project and the Prince of Wales is seeking to raise £10 to £15 billion pounds from private and public sectors. The private sector money is likely to be through a Rainforest Bond rather than donations. You can get an update and express your support at www.rainforestsos.org

The future of forests in a changing climate is uncertain. The changes already seen in the climate over the past half-century have altered how forests work and will have increasing effects on them in the future. The ability of forests to capture carbon is at risk of being lost entirely unless current carbon emissions are reduced substantially. Higher temperatures provoking more forest fires would result in the release of huge quantities of carbon to the atmosphere, further exacerbating climate change. The worst scenario predicts that the Amazon rainforests will spontaneously combust and turn to savannah grassland.

Other scenarios paint a more mixed picture. The concentration of carbon dioxide in the atmosphere directly affects the growth of trees. Surprisingly, current concentrations of CO_2 are not optimum for photosynthesis, so CO_2 emissions that raise the concentration in the atmosphere can be interpreted as a good thing because it can be expected to make the trees grow faster. Controlled environmental experiments on young trees typically show this to be the case and trees grow about 30 to 50% faster when the CO_2 concentration is doubled.[80] Mature trees are not thought to respond in quite such a marked way, as growth rate slows as the tree matures. There is also a catch. The experimental results are obtained when all other environmental conditions remain constant. This is not going to be the case. The higher CO_2 concentrations will also affect temperatures and rainfall. Warmer temperatures mean that many types of tree will come into bud earlier, as has already been seen in the UK and other places. This lengthens the growing season and although it would be expected to increase tree growth in most years, early budding also increases the probability of frost damage. A generally warmer climate will mean that trees may not be able to withstand a cold winter or that species which need cold weather for flowering or seed germination may die out. As an added complication, seed germination is also dependent on seed moisture content, which could be affected by climate change. However, the complexity of interactions between changing temperature, rainfall patterns and the timing of seed germination for individual species makes prediction of overall impacts practically impossible.

British trees are also likely to face problems with the changing patterns of rainfall that are expected. Drier summers and possible droughts will place stress on the trees; wetter winters will waterlog their roots causing some of the fine roots to die off. This in turn will make them less stable and less able to draw up water in the summer. Being less stable, they will also be vulnerable to being knocked down in high equinox winds. Clearly, trees, like humans, are going to be affected by climate change. However, as trees have the potential to be our friends, we should plant lots more of them. China has already recognized this and in 1998 started a scheme to replant millions of trees. China now plants more trees than anywhere else in the world. I would stress, however, that tree planting must occur in addition to the urgent need to cut our fossil fuel emissions.

There are three main ways you can promote tree planting beyond your own garden.

The first involves supporting the REDD initiative mentioned above. The second is to give money to a charitable cause that specialises in tree planting. The Woodland Trust is a UK charity that seeks to prevent further loss of ancient woodland in Britain. It also works to restore and improve the range of species living in woodland and expand the area of new native woods. These can be found in a variety of locations around the country and people are actively encouraged to visit them to increase their understanding and enjoyment of woods. You can also donate money to provide for tree planting in your chosen area. This latter activity can even be done using Tesco Clubcard Reward Vouchers, so you can use your shopping rewards to plant trees! To donate directly go to http://tinyurl.com/wtdonate (tinyurl is a website that allows you to shorten long web addresses into something easier to manage, so when you type that address you will be taken to the donation page of The Woodland Trust). You can also join The Woodland Trust as a member and/or get involved with them in various campaigns. For example, they still need help to plant about 8 million trees under their 'Trees for All' campaign. If you prefer to support the rainforests or want to help them as well, there are a number of charities to choose from. Visit www.guidestar.org.uk for a comprehensive guide to UK charities, their objectives and financial performance and you can even donate online once you have chosen your preferred organisation.

The third alternative is to invest in forestry directly. The following does not constitute investment advice, which you should seek from an appropriately qualified individual, but you might like to look at the Cochabamba Project.[81]

The Cochabamba Project Limited has been registered as an industrial provident society in the UK with the stated purpose of supporting a unique reforestation project involving smallholders in the Cochabamba region of the Bolivian Amazon.

Industrial provident societies conduct their business as cooperatives for the benefit of a particular community and each member gets one vote regardless of the amount of money they invest (which is limited to a maximum amount). Using this structure provides collective security, democratic control and a route for investors to get their money out. Rather than having to find another buyer, investors are able to sell their holding to the society after a qualifying time. Investors also do not have to wait for decades until the trees actually mature in order to receive a fair return on their investment.

The project itself has the objective of completing a reforestation project by working with 1,500 smallholders in and around the area of Cochabamba in Bolivia. It is a partnership between the farmers and investors with the farmers providing land and labour while the investors buy the seedlings and materials for them. The eventual revenue is then shared equally between the farmers and investors. The project has been given a gold rating by the Climate, Community and Biodiversity Alliance and although it should be seen as primarily a social investment, it hopes to return 7.5% p.a. investor returns.

There are, of course, lots of other potential commercial investments that you can make in forestry if you want to investigate them.

Recruit a Friend or Neighbour to the Cause
(The Labour of The Amazon Queen's Girdle)

Of all the stories of Hercules's Labours, there seems to more variation in the accounts of how he obtained the Amazon Queen's girdle than any other. The Amazons were a pretty fearsome matriarchal society in which men had to perform all the household tasks and women were famous warriors. Hercules set out by ship and after he had reached the Amazon lands, he was visited by Queen Hippolyte. Impressed by Hercules' muscular body and lion-skin cape (the one he had taken from the Nemean Lion), she offered him the girdle as a love gift.

That's one version, but others have Hera appearing as a troublemaker and provoking a fight during which the Amazon warriors charge Hercules whilst the Queen is on board his ship. Hercules, suspecting treachery, offhandedly kills Hippolyte, takes the girdle and proceeds to destroy the Amazonian army. Other versions are equally bloody.

A variety of versions of a story occurs frequently whenever a group of people get together, whether they are friends or not. Consider the type of conversation that goes on in many bars on a Saturday night as the day's sporting events are dissected. Or, at a more serious level, how the problem of trying to figure out 'what really happened' is part of the daily routine in our law courts in which an elaborate web of versions and connections are sifted through in order to be appraised by a jury of peers. A very large part of our lives is concerned with jostling to see whose story of events becomes the accepted one and a lot of our energy is devoted to trying to ensure that it is our own.

What I rather like about the gifting version of the story, however, is the idea that one's expectation of a hostile reception can often turn out to be a false assumption, and so it is when we interact with each other. When we are dealing with other people, we never really know what kind of reaction we are going to get from them, although many expect the worst (do you ever find yourself thinking 'I couldn't possible ask her that!' or 'there is no way he will agree to that'?). This often makes us reluctant to try to persuade people to our point of view. Yet when it comes to climate change, we desperately need to build such an enormous swing in the mainstream perception of the need for big actions, that this is exactly what you need to do. Remember it is thirty years since the first scientific warnings were given about the dangers of climate change, yet a mainstream understanding of the dire nature of the problem and the myriad of solutions available is simply not there. This chapter is intended to give you some ideas that will help you get to the position where you are confident to ask people for support and are in a position where they will gladly give it.

There is an odd tension about human relationships in that although we are inherently social creatures, we care rather more for our own affairs and rather less for those of others than we may readily admit. This can lead to some quite interesting situations when people get into discussions and each only advances their own point of view. If they eventually concur, then they part company with each thinking what a wonderful person the other is, so wise and intelligent. If their opinions diverge, then the tendency is for each to look to bolster his or her own position. Each will bring out all the logical and perhaps the emotional reasons why their point of view is better. They might even bring out some statistics or quote the work of eminent people who hold a similar view. Each side tends to get increasingly entrenched in their own position and the salvos of attack and counter-attack grow more voluminous. In other words, they have an argument.

Arguments are very difficult things to win and more often than not end when one party simply walks away or cedes some temporary agreement to let the other side think they have won. The big problem with winning an argument is that winners produce losers and people do not like the feeling of being defeated. So even those arguments that you may have won are unlikely to bring about lasting change as the loser will simply look for the next opportunity to revert to their previous position (this is why democratic societies use votes to bring debates to a conclusion and have a legal system in place to make sure people keep to whatever the agreement was.) There is a way of avoiding arguments altogether and increasing the chances of persuading people to come round to your point of view and that is, simply, to put yourself into the other person's shoes.

I noted in the introduction that we each experience the world in our own way and spending some time trying to imagine how other people think can pay great dividends when it comes to communicating with them. The key word in the last sentence was 'imagine': remember that you still don't know. That is why it is important to ask questions of others so you can better understand what their key concerns are and how they see things working. In doing this, you need to keep an open mind.

When the person you are talking to says something you

disagree with, the natural tendency is to think 'ha! You're wrong. Now let me put you right.' If you act on that thought, you are most likely to end up exactly where you were trying to avoid: an argument. A better approach is to say something along the lines of: 'That's interesting. Tell me a bit more about why you say that.' This should help you to discern their reference points and give you a chance to see how strongly held their view is. Remember, however, that you are still trying to reach a sincere understanding of the other person's position so try to empathise with their viewpoints and be respectful of their sources of information.

The object of this phase of the discussion is to try and find areas where you might have some common agreement rather than build up an arsenal of points you want to attack on. Keep the approach friendly, courteous and sympathetic. If you are succeeding in this phase of the conversation, you should be hearing the other person saying 'yes' to you in response to your questions.

Once you think you have a reasonable understanding of the other person's point of view, having established some common ground and a reasonable level of mutual respect, you can start to try to persuade them to your way of thinking. There are four basic approaches to persuasion and each one tends to work better with certain personality types, but be prepared to try each tack if the one you deploy first doesn't make headway.

The first approach is to try to get the other person to think the solution was their idea. It is actually quite a skilful thing to be able to pull off, but if you manage to master it, you will find that people not only agree with your point of view but also think you are a wonderful person for holding it. If you have been listening carefully to the other person's wishes, wants and concerns, you should be in the position to make suggestions for them to consider and build on for themselves. Once they have come up with *their* idea, it is important that you credit them for it and repeat back what a good idea you think it is.

To give you an idea of this in action, a few years ago my family were keen to get a dog. I was reluctant as I could guess who would end up doing most of the walking and feeding once the kids had got over the initial excitement, but we all trooped along to the Battersea Dogs and Cats Home. When we went for our interview, bringing our list of candidate dogs, the staff member told us that she wanted to introduce us to Fred, who had only been put up for rehousing that day, so we hadn't seen him in the public kennels. He was a '96% match' for us, she assured us. When she brought him into the meeting room, he bounded all over the place with a huge amount of energy and promptly did his business on the floor. A very good sign, apparently, as by not fouling his own kennel he was displaying the fact that he was housetrained! My wife and children took to him immediately but I was still pretty reluctant. 'Can we change his name?', my wife asked and on being told yes, started to solicit suggestions. The kids started throwing out lots of ideas, which my wife 'mmmm'd' to, until I finally suggested Pilot. My brother had a dog named Sailor and it seemed like an amusing family connection we could make. 'Yes, yes. What a brilliant name' said my wife. I now felt differently because I had named him and twenty minutes later we had completed the paperwork and took Pilot home with us.

Another way to persuade people and one which is currently used in a lot of climate change related advertising is to appeal to their nobler principles. This is the 'don't do it for me, do it for the kids' or 'don't do it for me, do it for the world's neediest' type of message. This kind of altruistic appeal clearly works: charities would simply be unable to function if it did not and altruism is a fundamental part of the human experience. However, you do need to use this method with care and consideration. Remember that, with persuasion, what you are trying to do is to show people that they're responsible for their own success and happiness. Repeated appeals to a person's altruistic nature that are not backed by some evidence that what you propose is having an impact can lead people to doubt that their previous support has been worthwhile.

It is also important to appeal to the nobler principles of the other person rather than try to portray yourself as being of noble intent. The fossil fuel companies are extremely keen on spending advertising money to proclaim their concerns for the future of society and to trumpet the efforts they are making to bring about a better future, whilst in reality they are spending more money on the advertising than the research. People do eventually see through such ploys, however.

A third method of persuasion, which also needs to be handled with care, is to dramatise your ideas in such a way that the recipient looks at things in a different way. For this to work effectively, you really need to understand your audience. It is a fine line between provoking a 'wow, I never thought of it that way before' and 'what on earth are you going on about?' Have a look at an evening's worth of television commercials if you would like to see what I mean. How many get the balance right for you?

The last method to persuade in this far from comprehensive list is to throw down a challenge. This can be very powerful and popular as it plugs into people's desire to excel, to feel important and to *win*. Challenges and 'playing the game' are fundamental to our societies, whether it is at work, in your neighbourhood or on the sports field. These concepts are frequently used as motivational tools in business where targets, budgets and entering new markets are frequently cited as objectives and goals.

The idea of setting goals and challenging oneself is primarily what underlies the self-improvement industry. Goals and challenges are great motivators, but need to be handled with care so that they are exacting enough to push people forward, but are not so difficult as to be unachievable nor so easily attained as to provoke ridicule. Having goals is not a new idea: Aristotle advised people to 'have a definite, clear, practical ideal – a goal, an objective' and this imperative of being specific is often a feature in advice on setting goals. Sometimes, however, you can be too specific. In the early part of the 21st century, General Motors, the American car manufacturer, set itself a goal that was stated simply as the number 29. This was the share of the number of the number of cars sold in the US that it had enjoyed in 1999 and it wanted to get back there. It became a real focus for the company's efforts and a mantra for everyday decisions. Sales managers tend to love volume sales targets because you can usually make them by cutting a deal and that is just what they did. GM never did meet its volume goal and by 2009 was close to bankruptcy.

An alternative approach was espoused by Leo Burnett, who founded the advertising agency that still carries his name, when he said: 'When you reach for the stars you may not quite get one, but you won't come up with a handful of mud either'. This may seem like more of a philosophy than a goal, but then we are back in the realm of arguing about how Hercules really got hold of that girdle.

This has been a short run through what is possibly one of the most talked about and written about facets of the human condition: how to persuade each other to willing action. If you want to delve deeper into this subject, just pick up any book on sales techniques or even Dale Carnegie's *How to Win Friends and Influence People*, which, even though it dates from the 1930s, contains ideas that pop up repeatedly in more recent material.

The key point is that the task in front of us is not just to make people *aware* of climate change, but to convince them to personally *take action* to mitigate it. There follows a few of the more typical stances you might encounter when talking with people, along with a suggested response that I have previously found some success with. Feel free to develop your own tactics and also to share them with others through the website that accompanies this book.

Some typical responses to climate change

'It's not going to affect me in my life time'.
'How can you be sure? There might be talk about 2030 or 2050 as the dates by which we will need to take action, but already an ordinary ship has sailed the fabled North West Passage between the Atlantic and Pacific oceans for the first time in known history. The Arctic Ocean may be completely free of summer ice within just 5 or 6 years. Don't you think that will have an impact? Don't you think you are going to live more than a few more years?'

'It is a problem our children will have to sort out
'That is certainly true. The idea that future generations have to cope with the damage caused by the current one goes back to the Old Testament and beyond. What we do know about climate change is that it is affected by the concentrations of greenhouse gases in the atmosphere and that once there, the gas hangs around for a long time, 30 to 50 years. We have already passed the concentration level that was previously the highest in human history. Don't you think we should be cutting emissions now to reduce the size of the problem our children will have to cope with?'

'It doesn't matter what I do, I'm insignificant'
'It can be hard to believe that an individual is significant at a global scale. But think for a moment and you will easily come up with the names of many individuals who are. To be significant is one of the greatest desires of many people and we mostly achieve it through our actions rather than our inactions. By choosing to act, we enrich our lives. Do you ever vote in elections? Do you know where your biggest carbon impact is at the moment?'

'It doesn't matter what we do, it's the Americans and Chinese who have to change'
'You are absolutely right that Britain is a fairly small contributor to global emissions at about 2%, but we only have about 1% of the world's population. Did you ever catch your parents saying to you when you were a kid: 'do as I say, not as I do'? Did it not rankle? Only by making the changes ourselves will we earn the right to ask others to do likewise. Besides, China is acting on climate change and so is America, but they need to be encouraged to do even more. So, do you know where your biggest carbon impact is at the moment?'

'I don't believe human activity is causing it, the planet is vast'
'The planet is indeed a huge and very beautiful place, but it is not infinite. What happens when you feed the exhaust pipe from a car through a hose and into the passenger compartment? Yes, the people quite quickly suffocate. When you pull the hose from the window, you don't actually change the chemistry of what is happening, you simply change the concentrations. If the Earth was a tomato, the atmosphere would only be as thick as the skin. We have been burning fossil fuels for a couple of hundred years now

and the concentrations of the most dangerous gases is rising quickly. Don't you think we should switch the engine off before we suffocate? Do you know where your biggest carbon impact is at the moment?'

'They don't really know what's going to happen'
'None of us does, that's why you seldom meet a poor bookie. Even experts in their field make mistakes when predicting the future. But think of it this way: the most common predictions are for average global temperatures to rise by 2, 3 or more degrees. When the planet was an average of 3^0 C cooler the ice sheet extended south of London and New York. How do you think the world might look when the temperature rises? We can avoid that future if we start cutting carbon emissions now. Do you know where your biggest carbon impact is at the moment?'

'What about the Ice Age?'
The Ice Age occurred when the concentration of levels of CO_2 in the atmosphere fell and ended when they started to rise again. Throughout history, we have been able to record that CO_2 levels have fluctuated between 200 and 290 parts per million and the temperature has risen or fallen pretty much following the same pattern.[82] CO_2 levels have now risen to 390 parts per million, higher than at any time in the past 2 million years and are still going up. This will inevitably lead to temperatures that humans have never experienced before. We can avoid that if we start cutting carbon emissions now. Do you know where your biggest carbon impact is at the moment?'

'It's too late'
'It might be. Even some eminent environmentalists like James Lovelock think things have gone too far already. But

they may not have. Besides, when have you ever known the Brits to simply give up without a fight? Where do you think you'll start?'

Of course, you don't have to persuade everyone yourself. You might like to join some of the campaign groups that are already up and running. Even Hercules had some help when he tried to take the girdle. So there follows some of the organisations you can get involved with. I have listed some more general ones, plus some specific groups for the main areas of food, transport and power generation.

http://350.org
This is a global pressure group with a single aim: getting world leaders to agree to reduce the concentration of CO_2 in the atmosphere to 350 parts per million (ppm). This is the level that has been identified by climate scientists as being the *upper limit* that will prevent dangerous climate change. Note the words in italics. The atmospheric concentrations of CO_2 have already passed this level and some lobbyists are claiming that the target should be 300 ppm. As we shall see later the political debate is talking about allowing concentration levels far higher and as 350.org is up and running and gathering support it is a good rallying point. Rajendra Pachauri who gave Agence France-Presse his opinion about the target in an interview said:

'As chairman of the Intergovernmental Panel on Climate Change, I cannot take a position because we do not make recommendations. But as a human being I am fully supportive of that goal. What is happening, and what is likely to happen, convinces me that the world must be really ambitious and very determined at moving towards a 350 target.'

Current concentrations levels of CO_2 are at 390ppm, a

figure that rises to 460ppm when the equivalent effect of methane is added. Given that the Copenhagen agreement didn't even mention concentration levels, there is a great opportunity to press the government to adopt this as the global target. You can add your signature online at 350.org

http://climatesafety.org
This is a spinoff website from the Public Interest Research Centre, an independent charity that seeks to help the wider public understand the latest research on climate change, energy and economics in order to increase its impact. It recognises that the science is running to keep up with reality and politics is still running to keep up with the science. There is a free report you can download from their website, which is a superbly clear summary of where the current scientific thinking is at.

www.stopclimatechaos.org
This claims to be the UK's largest group of people dedicated to action on climate change and to limiting its impact on the world's poorest communities. The combined supporter base is more than 11 million people and spans over 100 organisations, from environment and development charities, to unions, to faith, community and women's groups. Again, it has a simple aim, which is to demand practical action from the UK government to prevent global warming rising beyond the 2^0 C danger threshold. The website is a useful source of campaigning tools and events.

Compassion in World Farming (www.ciwf.org.uk)
Compassion in World Farming was founded over 40 years ago in 1967 by a British farmer who became horrified by the development of modern, intensive factory farming and now

peacefully campaigns to end all cruel factory farming practices. As we saw in the chapter on food, vegetarianism is the best option from a carbon perspective, but if you really cannot do without your meat, it is only reasonable to treat the animals with some respect.

No New Coal (www.nonewcoal.org.uk)
Again a single issue campaign initially set up to fight the Kingsnorth development in Kent, which is currently suspended. See the chapters on engaging with industry and governments for more details on why coal is such a bad idea.

Campaign for Better Transport
(www.bettertransport.org.uk)
Started in 1973, the Campaign for Better Transport undertakes research and lobbies to create transport policies and programmes which give people better lives. Through national and local lobbying and public campaigns, they seek to make good transport ideas a reality and stop bad ones from going ahead. The 'Take Action' page on their website is full of simple-to-follow points, and if you sign up to their monthly email you can see the effect you are having.

Airport Watch (www.airportwatch.org.uk)
We saw in the Transport chapter how the costs of flying simply don't stack up at present. Airport Watch is opposed to the aggressive go-for-growth policy of aviation expansion that the UK is pursuing under the 'predict and provide' approach that will see additional runways and terminal capacity across the country. Airport Watch are opposed to any expansion of aviation and airports likely to damage the human or natural environment (which given the carbon intensity of flying should really be all of them!)

To reach Superhero status you need to sign up to support the initiatives and help them financially where applicable. How much do you need to give? Well, the UN set a goal 35 years ago for developed countries to contribute just 0.7% of their gross national product (essentially national income) to help developing countries. Just five countries have managed to meet this modest goal. If we take the figure as a guide and factor in an UK average income of £30,000, then this calculation would suggest a donation of £210 per year or, in round terms, £20 per month. The median income, which is the mid-point of the earnings range, in the UK is lower than the average by an astonishing £8,000, so please do calculate what 0.7% of your income is and use that as your guide figure rather than £20 per month. Spread it as you like across the organisations.

Shaping your Local Area
(The Labour of The Cattle of Geryon)

Geryon was reputedly the strongest man alive, but had been born with three heads, six hands and three bodies stuck together. He was the owner of a herd of red cattle of marvellous beauty and Hercules had to acquire them without demand or payment. Hercules managed to overcome the guard dog, the herdsman and even Geryon himself without too much trouble, but the real difficulties were in getting the cattle back to Eurystheus. Obstacles were continually put in his way. A gadfly bit the cattle, irritated them and scattered them so widely, it took the hero a year to retrieve them. Then there was a flood which raised the level of the river so high that Hercules could not cross with the cattle, but he simply piled stones into the river to make the water shallower.

Eventually, he reached the court of Eurystheus, and that would have been the end of the Labours except he was deemed to have had help with two of them, so Eurystheus set him two more. Multiple bodies, things scattered wide, always more hoops to jump through: these are familiar experiences to anyone who has ever tried to engage with their local community.

Quite why it is so difficult is hard to fathom. Certainly, the British seem to have a fairly ambivalent relationship with their local communities. We provide fewer elected representatives at a local level than our mainland neighbours: 1 councillor for every 2,730 residents in Britain, as opposed to 1 for every 120 in France, 1 per 420 in Germany and 1 per 600 in Spain. And, it seems we are less interested in who takes these positions: voter turnout at local elections is generally not much better than a third of the electorate as opposed to over two thirds in general elections.

There are various viewpoints about why this should be the case, but a fairly common one is that councillors are often perceived as ineffectual. The planning decisions they take get overruled by the National Inspectorate and having to follow national targets means they are often seen as administrators meeting central government targets rather than leading local communities. The truth is, when it comes to taking action on your local environment, your local council is much more powerful than you probably realise.

'Your local council' is also a term that confuses many as your council comes in many different guises. In some areas, there are two layers or tiers: a district council and a county council. In others, there is just one: a unitary authority. In London, each borough is a unitary authority, with the Greater London Authority, comprised of the Mayor and the Assembly, providing strategic, city-wide government on issues like transport. In more rural areas, there could be a town or parish council covering a much smaller area. Councils are also run in different ways. You might have a directly elected Mayor, for example, with a cabinet.

The more typical form is that there are council officers, who are full-time paid employees and are essentially there to manage the delivery of services. Then there are elected councillors: these are the people who set the priorities and resources that are going to be used for providing your services. Although paid an allowance, councillors are generally part-time. As they are elected, you get a regular chance to let them know if you think they are doing a good job or not. They will usually live in the area they are serving and are mostly keen to hear from and help the communities they serve. Most will run some form of community 'surgery' or open meeting, so residents can have the opportunity to talk with them. You will also be able to get contact details from the local council or various websites like www.writetothem.com.

The financing of local councils is complicated. Although they raise an amount of money from residents through council tax and from businesses through business rates, the majority of their income comes from central government. However, not every local authority gets the same amount of money per person, as the allocations are based on a complex calculation to encourage them to adhere to Treasury objectives whilst allowing them to deliver service levels which are both suited to their local residents and which contribute to shared government objectives. In other words, it is a political bun fight, as is any judgement on how well they spend the money.

After the election of the Labour government in 1997, huge numbers of targets were introduced to ensure that local services were being delivered to the government's satisfaction, although in late 2009, there was move to reduce the number of central government targets and increase the accountability of local councils to their communities.

The services that local governments are responsible for are those day-to-day services that make up what we consider civilised living. Education, fire and rescue, social services, waste collection and disposal, parks and recreation, most transport, roads and highways and, crucially, planning, are all local government responsibilities. When it comes to climate change, there are four powerful ways they can have an impact: planning, procurement, waste and recycling and the energy efficiency of their own buildings. Some are very active in using their powers to make a difference and six have been awarded what the Government considers 'beacon status' for others to seek to emulate. These six councils have offered to share their expertise and they can expect to be busy as there are 462 local authorities who aren't there yet. The six beacons are:

London Borough of Barking and Dagenham for their incorporation of sustainability principles into the major regeneration project of Barking Riverside plus their ability to work with private sector companies like Ford, who have a large factory in the area.

The City of London Corporation has managed to drive its own energy consumption down by 35 per cent over the last ten years. It is the first council in the UK to develop a comprehensive strategy for adapting to the impacts of climate change, perhaps because London is one of the major cities thought to be at risk of flooding as sea levels rise. They have also introduced some innovative projects such as the Green Roof programme in which roofs are covered in plants, which better insulates them and cools the surrounding atmosphere.

Eastleigh Borough Council for their efforts to address climate change, which are reflected in key services such as waste, housing and the management of green spaces.

Middlesbrough Council is achieving an annual 1 per cent reduction in CO_2 emissions and its approach has been adopted by partners in the Tees Valley. Middlesbrough has taken a lead in getting householders and communities involved. It has also taken steps to protect those most at risk from the impacts of climate change including the 'Heatwave' plan for vulnerable elderly people.

Woking Borough Council has been working on combating climate change since the 1990s and has achieved a 30% cut in CO_2 emissions over its 1990 level. This has been achieved by cutting energy consumption and switching to locally generated, sustainable energy along with high levels of community involvement.

Worcestershire County Council has also achieved beacon status although it is a bit difficult to see what they have actually done apart from various declarations, strategies, pledges and agreements. Maybe someone who lives there will let me know if there has been any real action?

The latter point on Worcestershire may seem a bit harsh, in fact I hope I am wrong about their activity level, but it does illustrate a very important point about local (and central) government: they have mastered the art of using words to convey the idea of action while obscuring what that action might actually be.

Sometimes good things come out of this and sometimes unintended things come out of it. My own borough, the London Borough of Merton, has some fame in the Green world for introducing what is known as 'the Merton rule'. This was developed in 2003 and states that: 'The council will encourage the energy efficient design of buildings and their layout and orientation on site. All new non residential developments above a threshold of 1,000sqm will be *expected* to incorporate renewable energy production equipment to provide at least 10% of predicted energy requirements.' (italics author's own). At the time of introduction, this was widely lauded and other councils gradually adopted similar policies, to such an extent that around half now have some variant of them.

The choice of the 10% figure is a strange one. For most buildings in the UK, their biggest use of energy is to heat (or cool) the building. In the case of heat, switching to wood fuel boilers is a simple and substantial move to sustainability and had the policymakers chosen the figure of 30% instead, this could have kick-started that industry. Instead the 10% level seemed to promote on-site renewable electricity generation, but even then it hasn't quite turned out as expected. Developers complained that when they were developing a new building speculatively, it was difficult to know what the eventual energy use of that building would be. This is especially true of industrial space. Also, the developer has to bear the brunt of the capital costs of the renewable energy systems, but it's the occupants who take on the benefits of lower energy bills (as well as ongoing maintenance costs). As developers see little opportunity to pass the cost of the renewables through higher prices to occupants when renting or selling a property, there is no incentive for the developer to install anything but the cheapest solution that will meet the criteria. Rather than being a boost to onsite renewable and improved building efficiency, it became a cap that had to be met as cheaply as possible. And, cheaper kit is more likely to break down but the building occupier is not required to fix it. As a result, buildings could easily fail to meet their 10% requirement in the long run, but as developers aren't obliged to report on the output from the measures installed no one will really know.

The Merton Rule is soon to be superseded by the 'Code for Sustainable Buildings'. This is a code that concerns itself with the carbon emissions of the building itself rather than simply its ability to generate energy on site. There are a number of levels within this code, with each one being more ambitious than the last. It tops at Level 6 for zero-carbon buildings by 2016. This may seem a good objective, but as we saw earlier, the reality is that zero emission homes are possible right now so one wonders why we need to wait.

Merton may well be about to repeat the experience. The 'core strategy' for the area, which will set the tone for development over the next 15 years, includes these statements on climate change policy:

'All minor and major development, including major refurbishment, will be required to demonstrate the following:

- How it makes effective use of resources and materials, minimises water use and CO_2 emissions;
- Use of the London Plan energy hierarchy concept;
- How it is sited and designed to withstand the long term impacts of climate change, particularly the effect of rising temperatures on mechanical cooling requirements;
- The adaptation of the building form and construction to make installation of sustainability measures viable. The onus will be on developers to robustly justify why full compliance with policy requirements is not viable.
Residential Development
- We will require all new development comprising the creation of new dwellings to meet the highest commercially viable level of Code for Sustainable Homes (or any subsequently adopted set of national sustainable construction standards). Viability is defined as an increase in cost of no greater than 3% of predicted unit sales price. We will calculate this using the Merton Carbon Code Costs Calculator (MC[3]).'

Two points bother me with this: paragraph d. clearly indicates developers can avoid 'full compliance' if they want to argue strenuously enough and paragraph e. places a bar so low that a developer will be able to demonstrate 'non-viability' just by shopping around for his quotes. In Britain, we expect buildings to last at least a generation and more likely three of four. Allowing investments to be made in buildings that are outdated in efficiency terms even before they are built is to lock that capital away for generations and is a wasted opportunity.

I have made this point in a submission to the council as part of their consultation process because the planning process is one of the key drivers of what your local environment will end up looking like. It is not an easy process to understand. A cursory glance at a typical planning application will reveal a highly detailed document with closely argued points that seem like minutiae. These are generally designed to pre-empt possible objections and often seek to relate the application to sections in the various planning laws. The planning laws are the basis on which you can object to a planning application and they work in apparently odd ways. Thus, for example, you cannot object to a proposed new building on the basis that it is ugly and will block a treasured view, but you could object on the basis that its size and bulk is out of keeping with its surroundings.

This is the type of system that allows (and maybe even encourages) people to declare how passionate they are about renewable energy, but that this simply isn't the right site for a wind turbine/ biomass generating station/ solar array. Few people realise that they can also *support* a planning application so keep scanning your local planning applications (normally published in the local paper and always available from the council) and write to support any that are clearly involved with the efforts to reduce emissions or to generate local renewable energy.

One of the common defences used by the politician

when they are accused of not acting quickly or boldly enough is that the public don't want it or that it will cost businesses money. The good news is that you can very easily change their mind on the first of these. You simply need to go and talk to them, but it is important that you talk to the right person. In the first instance, this would be your ward or local councillor. If you're not sure who they are, a phone call to the council or a visit to their website armed with your postcode should quickly reveal it. Next step is to phone them up and ask if you can see them. Don't be shy about this part, remember they are local people and are interested in improving the local area.

When you meet them, ask them what the level of GHG emissions in the borough is and what the council is doing to reduce them. If you get an answer with lots of 'cross-cutting themes', 'local partnerships', 'framework analysis' or other such waffle, you can be pretty sure that your council has put a fairly low priority on action. After all, the Climate Change Act of 2008 targets emissions reduction for 2050. Ask them if they have signed up to National Indicators 185 (CO_2 reduction from local authority operations), 186 (per capita reduction in CO_2 emissions in the LA area) or 188 (planning to adapt to climate change) and see if that provokes a more concrete response. If you get a reply that includes putting wind turbines on urban buildings or making the town plastic bag free, you can be pretty sure that your council hasn't really grasped the nature of the problem or how to make an impact.

Local authorities that want to take action have an enormous amount of help available to them. For example, they might like to sign up to use CRED. This is a tool developed by the University of East Anglia's (UEA) School of Environmental Sciences, which is regarded as one of the very best research and teaching institutions in the world for interdisciplinary environmental sciences (although that standing took a bit of a pounding in late November 2009, when hackers broke into their computers and stole vast quantities of correspondence, some of which was published and implied that researchers had been colluding on falsifying data and results. The incident was gleefully used by people opposed to action on climate change to 'prove' it was all a hoax, thereby misinterpreting the ambitions of a few to condemn an entire community.) CRED stands for Carbon Reduction Programme and it encourages communities to take real action to address Climate Change. The government has set a target of 60% reduction in carbon emissions by 2050. CRED believes that this timeframe is too long and is building a community that is committed to cutting their emissions of carbon dioxide to meet a target of 60% reduction by 2025. So far, authorities in Norwich, Suffolk, Essex, Camden, Birmingham, Chester and Flyde have signed up to use the CRED infrastructure to generate their own community carbon reduction programmes.

Finally, if you do get an answer that suggests a firm plan for insulating council buildings, or switching to bio-mass heating or combined heat and power schemes to cover the majority of their building stock, or increasing recycling rates to 50 or 60 per cent, or putting in place a sustainable procurement policy and making local transport by foot or bicycle much easier, then please congratulate them on a job well done and suggest that if they haven't already shared best practice and case studies on www.carbon-innovation.com, a community-style website that most local authorities subscribe to, they should do so.

Using local democracy can be a very powerful way to change the environment immediately around you and, once you get involved in lobbying the council, you may be pleasantly surprised by how effective small groups of concerned people can be.

If you are not finding that you are making the progress as quickly as you would like, you can stand for council yourself. Either join one of the main parties, which would help to reverse a long-term trend of decline in membership, or set yourself up as an independent candidate. If you follow the latter route, please be under no illusion about the amount of work you will be committing to, but you will be joining a growing trend of people who are concerned enough about how they are governed that they want to be involved. You can find support at the Independent Network (www.independentnetwork.org.uk).

The Independent Network is not a political party and does not impose any political views on the people it supports. They do, however, insist that all affiliates are non-racist and non-discriminatory and subscribe to decent standards in public life. This means they should believe in selflessness, integrity, objectivity, accountability, openness, honesty and leadership in addition to a commitment to represent constituents' interests. It seems odd to consider this Superhero-ism, but when viewed against the historic track record of many politicians and especially the 2009 expense scandal, it appears that it is.

Encourage a Company
(The Labour of The Lernaean Hydra)

The Lernaean Hydra was a many-headed beast with a dog-like body and breath so venomous it could destroy life. It had been summoned by Hera to menace Hercules. Hercules' second Labour was to destroy it. This he did by flushing it out into the open and grappling with it. But when he cut off one head, two or three sprouted in its place. Hercules' accomplice set fire to the grove where they were fighting and Hercules used a burning branch to cauterise the wounds when he severed a head, preventing more from growing. Once he had defeated the beast, he disembowelled it and dipped his arrows in the gall. From that time on, the least wound from one of Hercules' arrows would prove fatal.

I have chosen this tale as a link to talking about business because for many people, and particularly the environmental movement, business is an embodiment of evil that hides behind many brand names or 'heads'. However, when you take business on and win, you can use its strength as a powerful weapon. Businesses are absolutely dependent for their survival on the existence of a market for their product.

While they can be slow to see how markets are changing (and some simply cannot see such change at all, as the demise of IBM demonstrates. It was once the world's biggest corporation, when mainframe computers were the norm, it was completely thrashed when PCs become popular and survives now only as a consultancy business), once they do recognise a shift in customer sentiment, they can become a powerful agent to push change forward.

Governments are dependent on business to provide tax revenues for them and employment for the populace. Businesses are dependent on the populace to buy their products. The populace are dependent on business to provide their basic necessities as well as life's luxuries. We are inextricably connected and while other systems have been tried, it is the liberal democratic capitalist model that has been most resilient. By its nature, it is a responsive system taking direction from markets (i.e. our purchasing patterns) or popular votes. As individuals, we are simultaneously insignificant and all-powerful. Additionally, by telling government and businesses directly what we want, we can have a strong impact on how things turn out. This chapter gives some ideas for how you can influence business toward a low carbon way of working.

Businesses are not all the same. There are three sub-groups that generally have a different approach to running their businesses and this is likely to have an impact on any input they receive from outside.

The first is what are described as 'lifestyle' businesses. These are businesses that are dependent on a small number of people, the owner or owners, for their continued existence. How the business behaves, what it invests in and how it treats its customers are at the owner's sole discretion and whim. If the owner retires or sells it on, the operation of the business will, most likely, noticeably change. This style of business is extremely prevalent in the UK and if you approach one with a request, the response you get will be entirely dependent upon the predilection of the owner. It is worth doing, though.

Britain has about 4 million businesses (compared to about 25 million households), the vast majority of which are small or medium-sized (SME), which means they have less than 250 employees. The business sector is responsible for about half of the UK's emissions. Whilst large companies are being legally obliged to increase their energy efficiency, the SME sector isn't yet covered. A small industrial estate may have the same energy consumption as a small town, so energy efficiencies or switching to renewable supplies will make a proportionally bigger impact than the actions of a household.

The other two groups can both be characterised as entrepreneurial businesses. That is to say they have developed a way of working that is consistent and independent of their owners insofar as they don't need to be present to make sure that standards are being adhered to. There are systems, policies and managers in place to make sure that things work as they are intended to. They tend to be larger businesses and share the characteristic that regardless of where you encounter them, the level of service or product will be broadly similar. What makes the difference between the two parts of this group is ownership: it can either be open or closed.

Closed ownership is where the business is held by a small number of private shareholders or a family. There are a number of large companies that fall into this section, such as SC Johnson, the household cleaning products company, William Grant's whisky company and the Virgin Atlantic airline. When you approach these companies you can appeal to them as a customer, a member of a community in which they operate or maybe even as an employee, if that's what you are. However, they have no obligation to do anything other than listen to you. The owner still has the absolute right to do as they please so long as it is within the law.

Open ownership is where you have the opportunity, either through becoming a member or through the purchase of shares, to become a part-owner of the business. This is an exciting prospect because the managers have a legal duty to their shareholders, which doubles up your power as both a customer and as a part-owner. Of course, part-ownership puts you in a position only as powerful as your percentage holding, which will normally be small. However, you are at liberty to align with other shareholders and to represent yourself directly to the business management. This has become a route used by a number of activists in both the US and Europe and, even where the managers have tried to avoid the issue, it has been a successful route.

Consider the US lumber corporation Weyerhaeuser in 2005. A consortium of groups, including the Rainforest Action Network (RAN) and the United Steelworkers Union, had assembled a group of shareholders who collectively owned more than $400,000 in Weyerhaeuser stock. This seems like quite a lot, but as the company is valued at around $8 billion it is only a small fraction of a percent. The idea of the group was to attend the company's Annual General Meeting and address pointed questions to board members about the company's practices of recklessly logging endangered old-growth forests in North America, using wasteful tree-cutting procedures, and abuse of trust in the cutting of billions of dollars' worth of cedar forests on the Haida Nation tribe's lands in British Columbia.

The company became aware of their intention and Weyerhaeuser CEO Steve Rogel decided to bring in a large contingent of security guards and eliminated the custom of allowing shareholders to ask direct questions to the board members. Instead, shareholders were directed to write their questions on cards, which board members then chose from.

From the company's perspective, the meeting did not go well. Many shareholders and representatives were furious that they were unable to communicate directly with the board. The steelworkers said they submitted more than fifteen questions in the meeting and got just one response. Some of the attendees began shouting and were forcibly removed from the meeting. RAN's Old Growth Campaign director Brant Olson said that Weyerhaeuser's elimination of direct question/answer sessions in the meetings 'set a dangerous precedent for other publicly-owned company meetings. It also shows the contempt Weyerhaeuser has for its shareholders, the people who actually own the company.'

Steve Rogel told the press after the meeting: 'Obviously there are elements of society who will do anything to have their views heard.' Yet it was Weyerhaeuser's attempt to stifle the activists that really sent a message that day. News about the meeting spread throughout the local and national press. That day, the *New York Times* ran a story under the heading: 'Manager to Owners: Shut Up'.[83]

The concept of corporate social responsibility has gathered pace in the past 5 years, but it is an area that is as complicated as business itself and covers corruption, human rights, interaction with community, workforce diversity and employee relations as well as environmental impact. Within environmental impact, companies are variously judged by their stance on water use, resource management, waste and recycling and greenhouse gas emissions. There are a number of schemes, awards, reporting initiatives and league tables that are produced by a myriad of organisations that makes it difficult to work out just how much good is actually being done.

One list produced by Business Ethics, a website that started life as a magazine in 1987 'to serve that growing community of professionals and individuals striving to work and invest in responsible ways', scores US companies' performance across eight criteria, one of which is environment. The latest list available on the site was for 2007 and of the '100 Best Corporate Citizens 3 of them actually managed a negative score on environment and 26 others had a minimally low score.

Unfortunately, the site is not entirely clear about how it judges companies (I guess because it doesn't want others copying its methodology?), but compare the approach to that of Dow Chemicals. Dow is the largest chemical company in the world and the 42nd largest corporation overall.[84] They have signed up to the Global Reporting Initiative, a scheme endorsed by the United Nations to provide a method of reporting on the sustainability of business practice being used. At first sight, it looks like just another piece of management speak. For example, here is how they describe what they do:

'Overall Vision
The articulation of Dow's vision and strategy with respect to sustainability is no different than the overall vision and mission of the Company:

Vision: "To be the largest, most profitable, most respected chemical company in the world."

Mission: "To constantly improve what is essential to human progress by mastering science and technology."

Inspired by the Human Element, we strive to constantly improve those things essential to human progress. From the clothes we wear to the food we eat. From the homes we live into the furnishings, fixtures and fittings that adorn them. Equipment that purifies water and materials that save energy. Products that make our daily lives easier, healthier, safer or more enjoyable.

Dow's chemistry has long played an integral role in keeping pace with society's ever-changing ambitions and aspirations.' [85]

Well, they already are the largest so that doesn't seem like much of a vision, but the really interesting part comes further into the report when it starts to talk about the environment. It is also written in management speak so is a bit of a struggle to get through, but stay with it because there is a translation at the end:

'Providing humanity with a sustainable energy supply while addressing climate change is the most urgent environmental issue our society faces.

Dow operates at the nexus between energy and all the manufacturing that occurs in the world today. Over 90 percent of the products made have some level of chemistry in them, so no one has more at stake in the solution – or more of an ability to have an impact on – the overlapping issues of energy supply and climate change than we do.

As a world leader in chemistry, Dow is uniquely positioned to continue to provide innovations that lead to energy alternatives, less carbon intensive raw material sources, and other solutions not yet imagined.

In fact, our science and technology has been contributing solutions to the global climate change and energy challenges since 1990. Our science has led to the development of alternative energy sources such as biofuels, photovoltaics and wind. Our products contribute to reduced energy consumption.

Dow's vision on overall sustainability is reflected in our 2015 Sustainability Goals – a public commitment that we hold ourselves fearlessly accountable in the pursuit of solutions that address climate change, energy and other pressing world challenges.

We commit to measure and report progress in the following areas:

- We will leverage the strength of the human element in our laboratories around the world and make unprecedented financial investments in R&D to achieve breakthrough solutions that will slow, stop and reverse global warming.
- Dow will advocate for an international framework that establishes clear pathways to slow, stop, and reverse emissions by all major carbon dioxide-emitting countries.
- Dow will advocate for and participate in the monetization of carbon in fair marketplaces, a critical objective in establishing country market mechanisms for cost-effective carbon management. Each country should be allowed to establish their own systems with targets set fairly for each industry sector.
- Wherever we operate, we are enabled by the energy and feedstocks available to that country through its own governmental policies. We will advocate for governmental policies that generate the most energy efficient and least GHG intensive processes and products possible. Further, Dow pledges to be the most effective and efficient producer using available energy and feedstocks, wherever we operate.
- Dow will continue to advance and bring its world-class know-how and expertise in energy efficiency and conservation to other companies and countries that are earlier in the technology cycle in order to deliver more rapid progress in reducing the world's GHG emissions.

- We will continue to focus R&D and engineering resources on improving yields and the energy efficiency of our processes. This will enable Dow to attain even lower energy intensity targets, and we will be recognized as the efficiency leader.'

This seems fantastic. A major corporation, and a US one to boot, openly supporting action against climate change. Their policy appears to saying something like 'climate change is the biggest thing we face and we are going to use all our strength to face it'. But then comes the wriggle.

The 2008 report states: 'As part of our goal in the area of climate change, we commit to reduce our GHG emissions intensity by 2.5 percent per year through 2015 from a 2005 base. By 2025 we will stop the growth of absolute emissions of GHG within the Company.' In other words, the company wants to continue to grow, and while it may be able to reduce the greenhouse gas per unit of production, the total number of units produced will mean that their total emissions will continue to grow for the foreseeable future.

The bind that Dow finds themselves in is symptomatic of our economic system. The owners of Dow Chemical Company are a diverse bunch of individuals and institutions. The top ten institutional investors as of November 2009 were: Capital World Investors, Dodge & Cox, Barclays Global Investors Na/CA, Marsico Capital Management, The Vanguard Group, Inc., Fidelity Management and Research Company, State Street Corporation, Brandes Investment Partners & Co., J.P. Morgan Investment Management Inc. and Apg All Pensions Group. Collectively, they own just over one third of the stock whilst a raft of other institutions hold another third of the stock. You might recognise one or two names on the list, but they are hardly household names. By and large, these institutions are investment funds, using someone else's money to buy these shares and are committed to managing it sensibly and to maximising the return on the investment. Ultimately their holding might go toward paying for somebody's pension or contributing to the payout of an insurance claim or just making somebody feel richer. The point is that the management of the company, who are charged with acting in the best interests of the company (which normally means maximising financial return over a period of time), are responsible to a group of institutions that likewise have a responsibility to *their* owners, who will in turn be a mix of institutions and individual owners. Nobody owns enough to be considered as 'the owner', so the managers are left to take the responsibility for the future direction of the company and the 'owners' instead tend to consider themselves as investors who simply hold the stock on a transient basis as long as its value continues to rise.

The price of the stock reflects the investor's confidence that the managers will be able to keep to their side of the bargain. Therefore, a rising stock price simply indicates a growing confidence that they will continue to do so. But there is always pressure, felt by everybody, to make as much money as they can. Faced with the pressure to satisfy the growth requirements of the 'owners', as represented by the investors, the managers are faced with a difficult balancing act. They need to demonstrate that they are investing in the company sufficiently to meet the needs of the future, whilst continuing to deliver appealing results for the current quarter or half year. No matter how strongly the managers as individuals might want to save the environment, they have to do so in the context of their financial performance.

In the case of the Dow Chemical Company, they have been very successful in pushing down injury and illness rates at the company (good for reducing cost as well as providing a safer environment). They have also reduced the number of miles that hazardous materials are transported (better efficiency of transport reduces cost as well as the likelihood of the material being involved in an accident), thus claiming an energy saving of 1,600 trillion btu (btu is a measure of energy from the Imperial system) over the past 14 years. If this seems to be a lot, it is: in the form of electricity, the same energy would exceed the 2007 annual output of all the UK's power stations. They even boast an overall reduction in their greenhouse gas emissions, with 2007 and 2008 showing a downward trend.[86] However, their stock has underperformed in both their sector and the wider market since 2006, and is only rated 'average' as an investment opportunity by Morning Star, a popular investment news site. The management can expect to receive pressure from investors to raise this level of performance, although not as much as they would if the rating fell to 2 or 3 stars.

I am assuming that you do not control sufficient funds to make an investment in Dow that would be significant enough to move the share price, but I'm sure the management would appreciate knowing that you support what they are trying to do. I always find that handwritten 'thank you' cards get noticed more than letters. You will find their address at the back of the book if you want it. You could also suggest that they shift their environmental vision from the margins to a place more central in their overall company vision in order to help speed their rate of change.

I have spent a bit of time on Dow Chemicals simply because it was a pleasant surprise to come across one of the expected 'bad boys' doing so much. They have joined with 31 other US corporations (all Fortune 500, including five of the top ten) and environmental organisations to form

USCAP, which calls on the US federal government to quickly enact strong national legislation to require significant reductions of greenhouse gas emissions.

The detail of the call is not as demanding as scientists would like nor as rapid in timescale, but nor is it a 'green wash'. They currently call for a stabilisation at 450 to 500 ppm with cuts of 60 to 80% by 2050, compared with stabilisation at 350 ppm with 80% cuts by 2020 called for by Earth Policy Institute and 350.org amongst others. But it is a refreshing change from the activities of the American Petroleum Institute, which during the summer of 2009 was reported to be organising rallies against President Obama's energy and climate change bill. This brings us to the most problematic companies: the coal, oil and gas companies.

Money, they say, is power, and the oil and gas companies have it in buckets. The largest corporation in the world is Royal Dutch Shell, an Anglo/Dutch oil company that by itself supplies 3.2% of the world's energy. Seven of the top ten largest companies in the world are oil companies. The revenue of the world's top ten oil companies is equivalent to the annual GDP of Britain. One year of profits from them would pay for the London 2012 Olympics ten times over, but could only provide Fred Goodwin-style pension pots to about half of the competitors!

This is a very powerful and rich sector. It is also one that is facing extinction. No one knows for sure how much oil is left in the world, partially because the Middle Eastern countries regard the data on their oil reserves as a state secret. Estimates vary from 50 to 100 years of supply. In commercial terms, that is beyond the planning requirement of good management. A company will normally plan in great detail what it is going to do in the next year, have a fairly detailed strategic plan for the next 5 years and an outlined strategic intent for the next 10 years. When your performance is measured in quarters and years, the future is a very uncertain place. That probably explains why Shell is not yet that interested in renewable energy. According to their website, their strategy is:

'To invest in large, integrated projects that will produce oil and gas for decades and will benefit both the countries holding these resources and countries dependent on oil and gas imports. These giant projects take a long time and a lot of money to build. They cannot be stopped or started quickly in response to short-term changes in energy prices or costs.

Competition for access to oil and natural gas resources will remain intense. We believe we can differentiate ourselves through our technology, our operational excellence and our ability to manage these complex and difficult projects in socially and environmentally responsible ways.'

At first sight, Shell appears to be making an effort towards sustainability: they are members of USCAP, they have reduced their own carbon emissions, they signed up to the UN Global Compact with Business and they divert some of their funds toward social investment. However, the cut in emissions is not entirely as it seems. In 2008, Shell controlled or operated facilities emitting 75 million tonnes of greenhouse gases, (measured on a CO_2 equivalent basis), about 7 million tonnes lower than the previous year, and approximately 30% below 1990 levels. However most of the difference between 2007 and 2008 was due to changes in their portfolio, as they sold a number of refineries. The emissions are still there, simply no longer owned by Shell. What's more, when Shell restate their emissions on the basis of the other companies they own or part-own, the total figure goes back up to 89 million tonnes.

Both figures seem small, however, when you consider

that the customers of Shell emit about nine times more by using Shell's products than Shell does in making them. The UN Global Compact is also a bit of a disappointment. It covers ten principles of conduct under the headings of human rights, labour, environment and anti-corruption. Surprisingly, in my view, it doesn't include any obligation or target for social investment which, given the global nature of large business and their predilection for setting up subsidiary companies in low tax regimes to avoid paying tax, is a major omission (before any Shell lawyers start getting wound up, this is not a specific allegation against Shell but large businesses in general, and it does not imply any illegality in their actions but simply highlights a common business practice that is of doubtful morality. The *Guardian* newspaper ran a major investigation of the practice and you can see details at http://www.guardian.co.uk/sustainability/corporate-responsibility-reporting.)

Shell does use some of its profits for social investment, but at a reported $148m in 2008 against profits of $26,277m it fails to reach the 0.7% benchmark that the UN sets for governments.

Shell's rather curious position on sustainability is further illustrated by a section on their website that says:

'Contributing to sustainable development is integral to who we are and what we do. So we talk regularly about sustainability in our engagements with investors, governments and communities and in our company advertising [...] It underlines the need to maintain economic growth and reduce poverty by providing more energy. That can lead to debate, since there are different views about what "sustainability" or "sustainable development" means, and about what an energy company's contribution should be. In 2008, one of our advertisements was challenged in the UK by an environmental NGO for calling several of our investments to meet growing energy demand "sustainable" that, in their view, were not. Their complaint was upheld by the UK Advertising Standards Authority.'

Upheld means that the Advertising Standards Authority agrees with the complainant, so the Shell advert was deemed to be misleading[87] but that seems to have passed by as simply part of "the talk". Perhaps the reality of Shell's position can be found in the section 'Shell energy scenarios to 2050' in which they describe two possible futures. One, which is called 'Scramble', envisages a future in which 'policymakers pay little attention to more efficient energy use until supplies are tight. Likewise, greenhouse gas emissions are not seriously addressed until there are major climate shocks'. In the second scenario 'Blueprints', 'the growth of local actions begin to address the challenges of economic development, energy security and environmental pollution. A price is applied to a critical mass of emissions giving a huge stimulus to the development of clean energy technologies, such as carbon dioxide capture and storage, and energy efficiency measures. The result is far lower carbon dioxide emissions.'

In 2008, Shell invested $2 billion in exploration for more oil, $3.1 billion in the Canada oil sands project and $1.2 billion in research and development of new technologies including Carbon Capture and Storage (CCS). CCS is a technology that captures the carbon dioxide from burning fossil fuel to prevent it reaching the atmosphere and storing it underground. It is a technology that has been under test for a number of years without yet producing a method that works on the scale required for commercial use. In some ways, it is rather like nuclear fusion, a technology that mimics what goes on in the Sun and which has the potential to provide unlimited, clean energy. Unfortunately, it was considered to be 20 years away from becoming viable when I was a boy in the 1970s and is still considered to be 20 years away from being viable now.

Renewable energy is not perfect as the storage issues still need to be solved, but at least it is clean energy and a mixed source strategy can ameliorate storage problems. Just like a gambler at the horse racing, energy companies cannot back every horse in the race and need to choose which horse to back. In the early part of 2009, Shell announced they were dropping all new investment in wind, solar and hydrogen energy. Shell, it seems, may be hoping for 'Blueprints' but its actions suggest they are part of the 'Scramble'.

BP is Britain's major oil company and was derided a few years ago for rebranding itself as 'Beyond Petroleum', something that was later quietly de-emphasised. But they do still regard themselves as holding high ideals. Their website seeks to attract good quality people to work with them because:

'All over the world we look for people who share our ambition to be competitive, successful and a force for good [...] the idea of being "a force for good" underlines all our activities worldwide. So the way we work is guided by values – integrity, honest dealing, treating everyone with respect and dignity, striving for mutual advantage, transparency and contributing to human progress.'

While they do believe that fossil fuels will continue to be the primary source of energy for decades to come, they are at least active in developing and marketing alternative energies. Probably not to the extent that many, including myself, would like but it counts as fairly substantial. In 2008, BP invested $1.4 billion in alternative energy and carbon capture and storage, compared with $15.6 billion for oil exploration. While the company explicitly recognises that the writing is on the wall for fossil fuels, they still see that wall far enough away to mean that the overwhelming focus of current management action is on finding and supplying more oil and gas. They recognise that a price on carbon emissions would change that and even urge governments to act locally ahead of global agreements if necessary but, as it stands, the business case is still beholden to oil.

So what can you do to influence the future behaviour of these companies? Ultimately the managers are responsible for the long-term health of the company and the shareholders hold them to account for their actions. The management will also listen to other so-called stakeholders: employees, suppliers, customers and members of the communities in which they have operations. Today's actions and tomorrow's plans will enact a balancing act between building shareholder value, thorough attracting more customers and improving operational efficiency, and avoiding shareholder loss through alienating potential customers, losing valuable staff or upsetting communities which risks potentially provoking punitive restrictions on their operations. The first thing to do is to decide which of these groups you belong to and clearly identify yourself as such when you write to them. You can become a shareholder very quickly and easily through buying some shares, either through your bank or one of the online share-dealing services that are available.

Being a shareholder will give you a powerful influencing position on companies, although obviously the strength of your voice is related to the number of shares you or your group have. From a legal perspective, shareholders have rights over and above any other stakeholder group, although sometimes it might not feel like it. Be warned, however: when you buy shares your money does not normally go directly to the company, but to the current owner of those shares. The current owner will be willing to part with them at a price but that price reflects his or her view of the future returns that might be gained from keeping them. The price is therefore a subjective evaluation rather than any real indicator of value. In just the same way as the price of bank shares dropped dramatically when it became obvious the trouble they were facing in the operation of their core business, the share price of oil companies may drop dramatically as it becomes obvious that the world is going to choose to move to renewable energy sooner rather than later. Therefore you may be putting your money at risk.

When lobbying, always write to a named director and, in regards to the key issues, write to each of them with the same information. The task is to make the point that the business risk from climate change is now apparent and will have an impact in the next 10 years. You can do this by referring to the Arctic ice melt which was analysed by the Catlin Arctic Ice Survey undertaken in March 2009[88] and also in the Climate Safety publication from the Public Interest Research Centre.[89] The disappearance of Arctic ice is happening at a much faster rate than predicted by the Intergovernmental Panel on Climate Change. Chart 4 earlier in the book would suggest it is perhaps as much as 60 years faster. Under these circumstances, it is a debatable assumption that fossil fuels will continue to be accepted as the main long-term source of energy, and a more prudent use of shareholder funds would be to invest in renewable energy development rather than oil exploration. Keep copies of all correspondence: it might just have the power of the Hydra's gall bladder one day.

Encourage a Government
(The Labour of Capturing Cerberus)

Cerberus was a fire-breathing giant feared by even the Olympian gods. The most common depiction of Cerberus in Greek mythology and art is as a creature with three heads and a snake's tail. Cerberus was always employed as Hades' loyal watchdog, and guarded the gates that granted access and exit to the underworld (also called Hades, or more commonly, Hell).

The task of capturing Cerberus alive was the final Labour of Hercules and was the most dangerous and difficult. After being given the task, Hercules was initiated in the Eleusinian Mysteries so he could learn how to enter and exit the underworld alive. He also recruited two guides to take him to the entrance. Whilst in the underworld, Hercules came across two of his friends, Theseus and Peiritihous, who had been captured by Hades and strapped to torture chairs. He managed to free Theseus but the earth shook when he attempted to liberate Peiritihous, so he had to leave him behind. After a journey through the underworld, Hercules met Hades and asked his permission to bring Cerberus to the surface, which Hades agreed to if Hercules could overpower the beast without using weapons. Hercules was able to overpower Cerberus by grabbing its three throats and holding them until Cerberus choked and yielded, although not without using his serpent's tail, from which Hercules was protected by the pelt of the lion he had acquired in the first Labour. Having subdued the beast, Hercules dragged it out of Hades and brought it to Eurystheus. The king was so frightened of the beast that he asked Hercules to return it to the underworld, after which he would release him from his Labours.

It is fitting to use the twelfth Labour when we turn to discussing politics because, in reality, politics is a difficult and occasionally dangerous realm.

In the first chapter, in my description of the immutable human characteristics, I noted how we each experience the world in our own way and how our reaction to our experiences is a big driver of our behaviour and that how we experience others and form ourselves into groups is critical to our survival.

We have a consciousness that is seemingly unique in the animal world. We are exquisitely and sometimes painfully aware of other's minds, particularly those of other humans. However, we can never know exactly what somebody else is thinking, we can only guess. We guess people's probable state of mind from their behaviour, particularly in regard to their intentions (he looks angry, I'm going to run), perceptions (on hearing someone's reply to something you said have you ever thought: 'you really didn't understand what I said, did you?') and their desires, but it is much more difficult to guess their thoughts, beliefs or knowledge. This is important because many of our most powerful emotions, such as pride, shame or embarrassment, are dependent upon what we guess others might be thinking about us. Because we are social animals, this can make people sensitive about what they reveal of themselves to others. After all, once a piece of information is known by others, it cannot become unknown especially in the highly interconnected age of the internet that we now live in.

The groupings into which we form ourselves are also

dynamic, and most of us have at some time suffered a shift of allegiance or a miscommunication that has turned a firm friendship into something altogether different. We experience the world in our own way and, for most of us, the world would be a much simpler place if only everybody else saw things as we do. That they don't once prompted the French philosopher Jean Paul Sartre to observe that 'hell is other people'. Hell is other people and politics is how we cope with that at the level of society.

Politics provides the framework on which everything else is built. It is the process by which society decides what it is going to do. Not everyone agrees upon what is right all of the time and so most countries have developed democracies to ensure that the will of the majority is the one that prevails. So far so rational, but another way to look at it is that we have outsourced our 'hell' to a small minority who are prepared to suffer it as a career choice.

Quite what motivates people to go into politics is probably as varied as the weather. Noble beliefs may be part of it for some and many politicians can be quite dogmatic as a consequence. Dogmatic beliefs can come from the political left or right or even the centre, and I am old enough to remember when the left and right were clearly associated with different parties. Today, quite a number of politicians have no consistent beliefs of any sort. Their aim is to win elections and be given a job and they're rather inclined to regard ideology as something that gets in the way of honest ambition.

What does probably differentiate people who become politicians is that they are very motivated by power. With power, of course, usually comes money, but this is where, in Britain at least, there seems to be a strange relationship between the governed and the governors. As a society, we pay our representatives what a corporation would pay a middle manager, someone who is responsible for making sure that product is nicely displayed on a shop shelf, or that heath and safety regulations are followed correctly. As a society, we reward our Prime Minister with the riches that would be easily obtained by the owner or manager of a moderately successful business employing around 25 people. The expenses scandal of 2009 clearly illustrated that many of the MPs found this level of reward rather desultory and found an over-generous if legal method of essentially topping it up. At the time of writing, the fury and public outcry that followed the expense scandal has yet to be vented fully at a general election but it soon will be. As 'villains' were to be found in all parties, the danger is that the covenant between the electors and the elected has been broken and that the turnout will fall disastrously as voters turn their back on politicians who are 'all the same'.

One recent politician who could hardly have been accused of being bland was Margaret Thatcher. First elected in 1979, she quoted Francis of Assisi on election night: 'Where there is discord, may we bring harmony. Where there is error, may we bring truth. Where there is doubt, may we bring faith. And where there is despair, may we bring hope'.

Depending on how you experience the world, she then set about what could be called either the greatest destruction or the greatest revolution of British life seen in centuries.

Using a particular brand of economic theory called monetarism and placing the individual rather than society or groupings at the centre, whole industries and geographies were transformed. Unemployment rose very sharply. Famously the definition of unemployment was changed multiple times to try and reduce the top-line figure. Conversely, those with jobs did rather well out of it.

Some new industries like IT appeared and banking enjoyed the start of its extended boom. Rather like the 1930s, those who did well did very well and those did badly did very badly. Political opinion over Thatcher became (and pretty much remains) extremely polarised, but she went on to win two more elections and her party won a further one.

The Conservatives eventually succumbed to New Labour in 1997, following prolonged allegations around 'sleaze'. Remarkably though, the winning party in those five elections only won between 42.3 and 43.9% share of the votes but their share of the seats varied from 51.6 to 63.6%. This was made possible by Britain's 'first past the post' electoral system which means that in each election rather than every vote being significant, only those in about 100 'swing' consistencies are really influential in determining the make-up of the government. This is important from two perspectives. Firstly, if the MP has a 'safe' seat, their concern for individual electors will have less of a bearing on their future than the fortunes of their party overall. The natural self-interest of the MP is to further their position in the party.

The party is influenced not simply by voters but also by their members, donors and other institutions that they interact with. There will be times when, like Hercules, the MP has to choose between saving a friend, the electorate, or their own vested interests. The electorate will not always be chosen.

The second important feature of the British system is that the ruling party rules, the losing party opposes. The system is not well geared to tackling urgent issues of common good simply because the opposition feels it is not doing its job unless it argues against the government. To agree is politically damaging. This means that the entire electorate is conditioned to see politics as an exercise in complaint.

The nature of climate change, which is all-pervasive,

needs a different approach. It is not a case of 'us' versus 'them', but a case of 'all of us' versus the laws of physics and chemistry. It certainly needs a global political solution because the alternative is a war solution. However, we need to make it very clear to our politicians what we expect of them and to express that in a positive way.

The first edition of this book was launched to coincide with the timing of the 2010 general election campaign. Wherever possible, the book was sent to the candidates of the three main parties in each of the 100 most marginal seats, so if you live in one of those listed in appendix 2 there is a chance your representative has read it.

Throughout the Labours, there are specific policy suggestions that can be used in specific areas, but there are some general themes that must be addressed at a national and international level. These are the leadership themes that will dictate whether or not we manage to make the transition to a sustainable world.

None of the major parties has adopted a manifesto as simple or radical as this. Please use the following policy list to challenge politicians to show leadership. By undertaking the Labours, we are being the change we want to see in the world. Only by acting with integrity do we earn the right to ask others to change also. There are only five policies that really matter and they are:

1. *The number is 350*

This refers to the concentration of carbon dioxide, CO_2, in the atmosphere.

During the whole course of human history until about 200 years ago, our atmosphere contained an amount of CO_2 that varied around a mean of 275 parts per million. Parts per million is simply a way of measuring the concentration of different gases, and refers to the ratio of the number of carbon dioxide molecules to all of the molecules in the atmosphere. 275 ppm CO_2 is a useful amount: without some CO_2 and other greenhouse gases that trap heat in our atmosphere, our planet would be too cold for humans to inhabit, but too much is clearly dangerous.

Since the Industrial Revolution and the start of our use of fossil fuel as the energy source of choice, the levels of CO_2 have risen and, by the end of 2008, had reached 385 parts per million. The global recession notwithstanding, that number continued to rise in 2009.[90] Current discussions speak of aiming to stabilise concentrations at anything from 450 to 700 ppm. That is far too high and puts the whole of civilisation in an unacceptable position of risk.

Coral reefs could start dissolving at an atmospheric CO_2 concentration of 450 to 500 ppm. Many scientists now believe the Arctic will be completely ice-free during summers from sometime between 2011 and 2015, some 80 years ahead of what scientists had predicted only a few years ago. The inhabitants of Cumbria experienced severe flooding in late 2009 following a storm that was reported to be a 'once in a thousand year' event, yet it was only four years since the last serious flooding in the region.

James Hansen of the US National Aeronautics and Space Administration, one of the first scientists to warn about global warming and who has been tirelessly doing so for more than two decades wrote: 'If humanity wishes to preserve a planet similar to that on which civilization developed and to which life on Earth is adapted, paleoclimate evidence and ongoing climate change suggest that CO_2 will need to be reduced from its current 385 ppm to *at most* 350 ppm.' (Italics added by author.)

This is a difficult thing to achieve but it is possible: it requires only that we switch from fossil fuel to renewable energy, stop deforestation and improve agricultural and forestry practices around the world. This will allow the earth's soils and forests to slowly cycle some of the excess carbon out of the atmosphere and eventually CO_2 concentrations will return to a safe level. The only barrier to achieving this is the vested interests of people who wish to continue with business as usual. But, to quote one of Margaret Thatcher's phrases when she was faced with opponents to her continuation of unpopular policy: 'There is no alternative.'

2. We must correct the accountancy error that put us here in the first place

Fossil fuel is all-pervasive in modern society, but its use started after the 'rules' of business and accounting had already been set. The desire of economists to ignore the extra costs of its use and treat them as 'externalities' has been the longest running exercise in sweeping the dirt beneath the carpet in history. Carbon has to carry a price that reflects its true cost to global society.

There are a number of ways this can be achieved: through putting a cap on total emissions and allowing the market to derive a price, or placing a tax on emissions, or using a fee and dividend scheme.

'Tax' has become a dirty word to many, especially in the UK and US over the past 30 years. In the UK, although the total tax take as a percentage of national income fell during the 1980s it has fluctuated around the 37 to 38% range for the past 18 years.[91] What has happened, however, has been a fairly radical change in the structure of taxation over that time (primarily away from income tax, especially for the better paid, towards indirect taxation and taxes on employment). There is therefore a good precedent for moving taxes around, even if there are still political challenges and potential outrage from vested interests (when the US administration floated the idea of taxing high-calorie fizzy drinks as a way of tacking obesity and diabetes, the food industry lobby groups indignantly declared that 'the tax code should not be used as a tool for social engineering nor should it be an instrument for penalising individuals' personal food choices').[92]

Given the pernicious and widespread nature of the damage from greenhouse gases, it seems to me to be entirely reasonable to use the tax system to discourage their use, just as it has been used to discourage tobacco use. Despite the cries from motoring lobbies in particular, the current tax take from fossil fuel is actually remarkably low: just 4.6% of government revenue, about comparable to Council Tax, and a fraction of the 27.7% from income tax or 17.9% from National Insurance.[93] There is a strong case to be made for using the tax system to encourage people away from fossil fuels. For example, VAT on domestic heat and power is currently lower than the standard rate. Raising the rate for fossil fuel generated to the standard rate while leaving renewable power at the reduced rate would give an immediate and significant boost to the rate of switchover. Another possibility would be to follow the German example and shift tax away from labour and towards fossil fuel energy. Since 1999, Germany has used this method to reduce their emissions and generated an estimated 250,000 jobs in green energy. Some 2,500 economists, including 9 Nobel Prize winners, have endorsed this strategy.[94]

The EU has also tried the first approach and introduced an emissions trading scheme in 2005. Unfortunately, the allowances have been too generous and following the economic recession prompted by the credit crunch, the price of carbon actually fell throughout 2009, and in November of that year was trading below $20 per tonne, well short of the $50 which the economist Nicholas Stern estimated was necessary to provoke the global economy into enough action to stabilise emissions at 550 ppm (and remember that 550 ppm is a very risky figure, the target is 350 ppm).

Stern's estimate of $50 per tonne is just that; an estimate. If we use data from the International Centre for Technology Assessment and quoted by Plan B[95] who calculated the total indirect costs of GHG to society to be equivalent to $3.17 per litre of petrol, then the cost of a tonne of carbon dioxide from the transport sector is $1,344!

The fee and dividend scheme is a carbon tax with 100 percent dividend. The tax is applied to oil, gas and coal at the mine or port of entry, and is promoted by James Hansen as being the fairest and most effective way to reduce emissions and foster the transition to the post fossil fuel era. It would ensure that unconventional fossil fuels, such as tar shale and tar sands, stay in the ground, unless an economic method of capturing their CO_2 is developed. The key to its success lies in the entire tax being returned to the public, equal shares on a per capita basis (half shares for children up to a maximum of two child-shares per family), deposited monthly into bank accounts. The cost of goods using fossil fuels would rise and the cost of renewable energy would stay the same. People would have more money in their pockets to choose products and services than they would otherwise, and may now find choosing renewable goods more attractive. As demand for renewable goods rises, they enjoy economies of scale, making them even more attractive. Demand for fossil fuels would start to decline and society's transition to a sustainable model would be achieved.

Whichever scheme is adopted, the important thing is that it promotes a rapid flight from carbon-based energy to renewable-based energy. The current cap and trade scheme certainly is not capable of that and to become so would need to reduce the size of the cap on a much more ambitious timetable than it presently has. To do that it will probably be necessary to remove the freedom of individual European Union member states to set National Allocation Plans and remove offsetting as a possibility. (Offsetting is the mechanism whereby it is deemed OK to keep producing greenhouse gas if you also fund a project elsewhere that will capture or reduce carbon. However, remember we are already above the concentration level that is now considered safe so all focus needs to be on replacing our dependence on fossil fuels in their entirety). After the banking crisis and the MP expenses scandal, there might be a lot of political capital to be gained by any party that adopts the fee and dividend route.

3. Just say 'no' to new coal

If the policy above is properly implemented this one won't be needed, as the cost of coal generation would be unattractive and all investment would go into renewables. As Britain has a poor record of implementing radical change, though, it is better to include it as a 'belt and braces' approach.

Coal is the most polluting form of fossil fuel and burning it doesn't just release greenhouse gases, it also produces large amounts of other poisonous substances such as mercury. Britain's biggest single emitter is the coal-fired power station at Drax, which is responsible for about 4% of the country's total.

Unfortunately, unlike oil, where total depletion can now be forecast within our lifetime, there are still quite large reserves of coal around the world, much of it in the developing world. The US Department of Energy still believes that 'coal is one of the true measures of the energy strength of the United States'[96] and the World Coal Institute believes there are sufficient supplies to last 122 years. Globally, the use of coal is the fastest growing fuel and, appallingly, more plants are planned in developed as well as developing countries.

Not only that, but many governments still subsidise the industry, particularly in the former Soviet Union and China, but even here in Britain it has received a minimum of £53 million over the past 6 years. Britain is also using more coal, with total production up 10% and imports up 13% in the first quarter of 2009 compared to 2008 and an increasing numbers of approvals being given to opencast coal mining projects in the country, with 54 new ones approved since 2005 alone.[97]

Perhaps it is with coal more than any other fuel that we need a major breakthrough in global thinking. Possibly the most effective example of global action to avert an ecological problem was the Montreal Protocol of 1987. This addressed the problem of the depletion in the ozone layer, particularly noticeable in the southern hemisphere, which put large populations at increased risk of cancer. The protocol mandated the phase-out of the use of chemicals called chlorofluorocarbons or CFC for short, which were commonly used in refrigeration and industrial cleaners.

One company particularly affected was a Canadian company called Northern Telecom who used a CFC to clean circuit boards after production. Following extensive but fruitless searches to find substitute chemicals, one engineer posed the question: 'how do these circuit boards get dirty in the first place?' This constituted a breakthrough in their way of looking at the problem and led to a redesign of the production process, which resulted in no cleaning being required and the subsequent production of better and less costly boards.[98] We need a similar breakthrough in our way of thinking about coal.

There is enough energy from the sun hitting the planet in one hour to power all human activity for a year.[99] At the moment, it is simply not as convenient or transportable as fossil fuels, but electricity is electricity. Professor David MacKay, in his excellent book *Sustainable Energy – without the hot air*[100,] demonstrates the practicality of using concentrated solar energy (CSE) power plants in North Africa to supply Europe with its electricity in its entirety. This is the type of bold plan that is needed. Instead, the coal and generating industry have managed to persuade governments that they have a magic solution for the future in the form of carbon capture and storage or sequestration (CCS). This technology was mentioned in the 'Encourage a Company' chapter. Nobody knows for sure whether it will be possible to make it work effectively or if the stored carbon dioxide will simply become as big a headache as spent nuclear fuel. Nor does anybody know how much it will cost to develop. The UK estimates for a single demonstration plant are between £750 million and £1.5 billion which equates to between one and two times the amount of money the government currently raises through the Climate Change Levy.

So, on the one hand, we have CSE, a technology that is truly sustainable and everlasting, but, on the other hand, an enormous gamble that the old technology can somehow be cleaned up with CCS. It is just like asking 'how did this thing get dirty in the first place?' as opposed to 'how do we find a different cleaner?'

4. Give investors a little TLC

Money does not flow around the world evenly, but rather tends to concentrate into the hands of a smallish number of individuals and companies. I say smallish because the number of billionaires is counted in the hundreds, the number of very large corporations is counted in thousands but the number with some form of investable, excess wealth is counted in the millions.

About 3 billion of the people on Earth have an income of under $1,000 per year, but the richest 10% of Americans, about 30 million people, have an average income in excess of $45,000 per year. In terms of assets, an estimated 8 million people globally have an average of $3.7 million each.

Within the richest, Bill Gates has an estimated $40 billion and Steve Jobs, the 100th ranked billionaire, has $3.4 billion. So Jobs has 10 times less than Gates, but nearly 1000 times more than the average. Even within the very rich there is great inequality.[101] However there are too many rich people to get into a room at any one time to force them to agree on how to sort out the world's problems, so we need a policy framework that will encourage them to invest in low and no carbon projects and activities.

Investors want to minimise risk and they calculate the risk level by considering the transparency, longevity and certainty (TLC) of government policy. Unfortunately, many governments around the world are not currently delivering policies of that standard, according to a Deutsche Bank report. Although Britain is better than, say Italy, it falls behind a group of major countries including Australia, Brazil, China, France, Germany and Japan.[102] In fact, the report concludes that:

'Even if current and select proposed policies were to make their maximum possible impact, emissions in 2020 would still exceed 450 ppm, the amount needed to limit the average world temperature increase to 2^0 C. To meet such a goal, emissions would need to be reduced further, by an amount equivalent to the current annual emissions of the U.S.

economy. More capital is required to mobilize climate change industries, and more action by government is required to attract capital. Investors are most attracted to countries and regions with comprehensive, integrated government plans that are supported by strong incentives, such as feed-in tariffs. Energy efficiency could help deliver significant reductions in emissions. Since efficiency provides savings in the long-term, it is essential that governments tackle market failures to encourage capital deployment in this area.'

In October 2009, a Bank of England report concluded that the governments of the US, UK and Eurozone countries had spent over $14,000 billion, about a quarter of global annual income, on propping up the banking system. As the destruction of assets, such as people's houses and infrastructure such as bridges and roads, will get increasingly worse as climate change continues to take hold, Britain and the rest of the world urgently need to put incentives in place which encourage investment in renewable energy and energy efficiency, and discourage investment in fossil fuels. To ensure that that oil companies develop wind and solar power rather than seek out oil and gas, and generating companies invest in solar and biomass rather than CCS, individuals are encouraged to install their own rooftop systems to generate heat and power and discouraged from using jet travel for foreign holidays. Funnily enough, all of these things would come about if the second policy, the setting of a price on carbon, was sufficiently effective. The total estimated amount required varies enormously and I have seen figures that range from $187 billion to $1,206 billion per year.[103] Although these are extremely large numbers, even the higher figure is less than 2% of global income, which seems like a small price to pay to protect civilisation as we know it. And it is a lot cheaper than saving the banking system.

5. Adopt a new measure of progress

As our lives are busy and complex, we like to try and simplify things to try to make them easier. We look for measures and signs to work out how things stand.

At a national level, the total level of economic activity is measured as the gross national product (GNP) and gross domestic product (GDP). The difference between the figures is mostly down to ownership. GNP doesn't include goods and services produced by foreign producers, but does include goods and services produced by domestic firms operating in foreign countries. GDP, on the other hand, includes only goods and services produced within the geographic boundaries of a country regardless of the producer's nationality. They are both statistics, which should in theory mean that they have a neutral value (statistics are, after all, simply numbers), but in our global industrial civilisation they are mostly interpreted to mean things are getting better (the measure goes up) or worse (the measure goes down).

The US politician Robert F. Kennedy's view of GNP was that it 'measures neither our wit nor our courage, neither our wisdom or our learning, neither our compassion or our devotion to our country. It measures everything, in short, except that which makes life worthwhile and it can tell us everything about America – except whether we are proud to be Americans.' Surprisingly, it has taken nearly 40 years for economists to develop a more balanced measure of how well society is doing.

An organisation called Redefining Progress is a US public policy think tank dedicated to sustainable economics. In 1995, they developed a Genuine Progress Indicator that includes all of the usual data on GNP, but then started adding or subtracting values for other items that contribute to how well a society is functioning. Primarily, it makes positive adjustments for factors such as income distribution (where international studies have found that narrower income distribution tends to promote general happiness in the population), the value of household and volunteer work, but makes negative adjustments for factors such as the costs of crime and pollution. Interestingly, it also includes a measure of dependence on foreign assets. The rationale here is that if a nation allows its capital stock to decline, or if it finances consumption out of borrowed capital, it is living beyond its means. The GPI counts net additions to the capital stock as contributions to well-being, and treats money borrowed from abroad as reductions. If the borrowed money is used for investment, the negative effects are cancelled out, but if the borrowed money is used to finance consumption, as has been the case in the US over recent years, this carries a negative score in the GPI. They have calculated data for the US that goes back to 1950 and it makes for interesting reading.

While GDP per person rose just over three times from just under $12,000 to nearly $37,000 per year, the Genuine Progress Indicator rose just 1.75 times and, since 1982, has fluctuated around the $15,000 per year mark. While the cost of air pollution has come down, the saving was more than cancelled out by increases in water and noise pollution. While the value of volunteering has risen more than fourfold, the rise has been more than negated by the rising cost of underemployment. The value of the services of labour-saving consumer durables has risen to 5.5 times its 1950 level, but any gain has been more than lost to increased commuting and loss of leisure time. Americans, it seems, are running faster to stand still.

The beauty of adopting a measure such as the Genuine Progress Indicator is that it allows us to continue progress. We earlier noted Adam Smith's observation about man's desire to improve his lot. This simply takes a better look at what improvement actually means. Its implementation is gaining supporters in China and Australia amongst other countries. Adopting it in Britain, or even at a global level, would be a great step forward.

In addition to developing the 'big picture' policies, there is the rather more mundane element of ensuring that the mechanisms are in place to guarantee that they are implemented. A plan is not an accomplishment. A policy is not an achievement if no one is there to implement it.

An example of this can be found in the UK's attempts to fight crime. In 2002, the Proceeds of Crime Act gave police and customs officers the power to seize money suspected to be the profit of crime or intended for use in crime and the amount involved can be as low as £1,000. They don't have to prove the owner of the money has actually committed any offence, as it is up to the owner to prove the legitimacy of the money, although the officers do need to get the approval of a magistrate within 48 hours. Despite the apparently low burden of proof required, the amounts of money seized have been remarkably low and in 2007 the Assets Recovery Agency, the body initially set up to lead the hunt against the criminals and now folded into the Serious Organised Crime Agency, was criticised by the National Audit Office for spending £65 million in order to recover £23 million. By the 2006/7 financial year, the amount had risen to £125 million, but as the estimated value of criminal activity is £18 billion, this represents 0.7% of the total. If this figure rings a bell, it should: it is the same as the target contribution for rich countries toward achieving the millennium development goals that respond to the world's main development challenges in eradicating poverty and promoting health and education. Maybe the government just got their targets mixed up?

Our government's record when it comes to taking ground-level action is a sorry one, with programmes being too difficult to understand, applied for short windows of time and subject to sudden change. Sometimes, their actions are just downright bizarre. A flagship initiative for 2010 is the introduction of a programme that was initially called the Carbon Reduction Commitment and was aimed at the country's largest commercial users of electricity. As one of the quickest routes for an electricity user to reduce their carbon output is to switch to renewable sources, you might have been tempted to think this would be a great boost. This is not the case here. There is already an incentive to generate electricity from renewable sources and the government didn't want to encourage double counting so that was excluded from the incentives list. Instead, they have changed the name and it is now called the CRC Energy Efficiency Scheme and seeks to reduce carbon emissions by reducing electricity consumption.

The introduction of so-called feed-in-tariffs (FITs) has been another challenge. FITs are designed to pay homeowners for generating electricity through installing technology such as a small wind turbine or solar cells in their property. They work very well because rather than having to find a few investors to make large investments in large-scale projects, you encourage lots of people to make small investments in small projects. Although the UK FIT for electricity was planned for April 2010 and for heat generation in April 2011, even as late as January 2010 the government department responsible for their introduction could not answer simple questions like how much people were going to get per unit generated or whether systems installed before the introduction of the scheme qualified for the benefit.

Getting better co-ordination and quicker decisions is an

absolute must. At present, the Government funds a number of initiatives like 'Act on CO_2' which is a communication platform sponsored by four government departments, and their website trumpets some of their successes. Asda and Tesco are featured as Partnership Case Studies to show what can be achieved. Meanwhile, government-funded watchdog Consumer Focus published a report called *Green to the Core*, which named Asda as the least green supermarket and claimed that Tesco has made no progress since the last report in 2007. Not too surprisingly, Asda and Tesco have reacted rather crossly, calling the report 'poor quality, misleading and inconsistent'. Actually, as Act on CO_2 has driving 5 miles less per week as a 'top tip' for averting climate change, maybe everyone needs to have a calmer look at where real impacts can be made.

Impacts can often be made when you have access to resources, so perhaps the biggest indicator of how far the Government has to go comes from looking at the budgets allocated to various activities designed to stimulate economic and carbon reduction activity. The Carbon Trust is a private company that has been set up and funded to accelerate the move to a low carbon economy, by working with organisations to reduce carbon emissions now and develop commercial low carbon technologies for the future. In its few years of operation, it has received a decreasing amount of funding from the government and is scheduled to receive £85 million in 2009/10. The Regional Development Agencies for England, which are charged with being the strategic drivers of regional economic activity, are scheduled to receive £2.2 billion pounds in the same period and will also manage £9 billion of European funds directed toward economic development. Search the websites of the Carbon Trust and you will find no mention of the Regional Development Agencies. Search the Regional

Development Agencies website and you will find scant evidence of The Carbon Trust. The lack of co-operation and integration, never mind the disparity in budgets, beggars belief.

So what can a Superhero do in the face of the enormity of government bureaucracy?

Your MP and MEP are obvious conduits, but it can also be helpful to use some of the other influencing mechanisms available. The government frequently holds consultations where anybody is welcome to express their view on matters of policy under development. In the area of climate change, the most important ones are undertaken by the Department of Energy and Climate Change, and you can see what is currently being consulted on at http://tinyurl.com/openconsult

In terms of campaigning, you can start petitions to the Prime Minister at http://petitions.number10.gov.uk/ and encourage as many people as possible to sign up. You can also encourage your MP to set up or sign an Early Day Motion (EDM). These are formal motions submitted for debate in the House of Commons, but very few of them are actually debated. Instead, they are used for reasons such as publicising the views of individual MPs, drawing attention to specific events or campaigns, and demonstrating the extent of parliamentary support for a particular cause or point of view. An MP can add their signature to an EDM to show their support. They can also submit amendments to an existing EDM. Although the majority of EDMs are never debated, there is a sub-section to challenge laws made by Ministers under powers deriving from Acts of Parliament (known as Statutory Instruments). Rather quaintly, these are known as 'prayers' and more of them make it to the debating chamber.

The key is to be consistent, persistent and positive. Tell

them what you want and when. 'Hell' may be other people, but remember that to them, you are the other. Your belief that climate change is a problem that you want to take action on, which was gained or strengthened in the first Labour, will sustain you. You have made the decision to make a positive choice for change. The power of that choice will be your lion's pelt. Hold the multiple throats of the politicians until they yield.

PART THREE

Where to from here?

The last time that society and science got into quite such a spat was probably in the 16th and 17th centuries, when theories about whether the Earth or the Sun was at the centre of the universe was a hot topic of debate.

Copernicus first put the latter theory forward just before his death in 1543, but it didn't gain much popularity until it was championed by Galileo in 1610. The Catholic Church, which at the time was fighting against the Reformation but was still a powerful influence in world affairs, rejected the theory. After Galileo advanced it again in 1632, he was forced to recant by the Church and endured being placed under house arrest for the rest of his life. It wasn't until 1758, some 148 years later, that the theory was finally accepted by the Church. In the meantime, the planets and suns carried on moving in the same way as they always had, regardless of what man or the Church believed.

Climate change is in a similar situation because at the heart of it is a physical and chemical system that dictates that the temperature of the planet will be determined by the amount of energy it absorbs from the sun minus the amount that it radiates back out. We have known since

1824 that the composition of the gases in the atmosphere affects the transfer of heat and the planet's ability to radiate it. By 1896, Arrhenius had worked out the impact of various amounts of carbon dioxide on temperatures. In 1908, he suggested that the emissions caused by human activity would be enough to influence these concentrations and provoke rising temperatures.

Business, particularly big business, has the power in today's society that the Church held in the 16th and 17th centuries, and it is business that is primarily behind the movement of denial.

The extent of their counter-activities has been truly astonishing and the profile of the anthropogenic/denial debate has risen greatly in the UK, US and other countries during 2009, both in terms of its quantity and emotion. As was mentioned earlier, in November 2009, computers belonging to the University of East Anglia were broken into by hackers who stole large amounts of correspondence. Climate change sceptics studied the emails and alleged they provided 'smoking gun' evidence that some climatologists had colluded in manipulating data to support the anthropogenic view.

The level of the debate is in serious danger of falling into a schoolyard spat between two bullies. This is a great pity because the element that has been missed in the 'debate: is: 'what happens if the other guy is right?' This is also known as the 'what is the worst that can happen?' factor. If the climate change deniers are right, but the world continues to invest money to switch away from fossil fuel, the worst that can happen is that some new industries are created about 50 to 100 years before 'economically necessary'. After all, these industries will need to develop as fossil fuels begin to run out, so all that we are left with is an industry ahead of its time and cleaner air in our cities. If the advocates of climate change science are right but the world does not massively invest *now* to move away from fossil fuels toward renewable sources of energy, then the whole fabric of civilisation itself is threatened. A small rise in average temperatures will provoke large changes in natural systems, particularly the release of vast amounts of greenhouse gases from the frozen northern tundra, and the whole process quickly escapes our control.

Some of the failure to make progress can be placed at the door of the Green movement, who have tended to advocate their view of how societies should live rather than simply discuss the fuel it should choose to power its activities.

I am conscious that many of the news reports I have quoted have come from the *Guardian* newspaper, and that in itself will provoke some readers to assign me to a certain 'tribe'. Tribes are an important aspect of the human experience. Although the human race emerged in Africa, we quickly split up, spread out and thereafter developed particular racial then tribal characteristics. Tribes are important because we are more powerful in groups than we can ever be alone. Our early history is the story of racial/tribal groups going to war for territory and the most fertile land. As the winners invariably interbred with the vanquished, the characteristics of the original tribes became blurred.

As progress continued, social tribes started to emerge in new hierarchies of power. It is possible to trace a genetic line from feudal lords to many of the richest families in Britain today, but they no longer hold the power they once did. Since the end of the World War I, the way in which power is acquired and relinquished has become increasingly more fluid. In today's complex and largely digital world, the power of social tribes is rapidly growing. Their common focus may have changed, but the strength of feeling amongst today's tribes is just as strong as it has always been. Consider this piece, which is actually a television review (and, yes, it is from the *Guardian*: it is from Charlie Brooker's 28 September 2009 column and is used here by permission of Guardian Media Group):

'Microsoft's grinning robots or the Brotherhood of the Mac. Which is worse?

Windows works for me. But I'd never recommend it to anybody else, ever.

I admit it: I'm a bigot. A hopeless bigot at that: I know my particular prejudice is absurd, but I just can't control it. It's Apple. I don't like Apple products. And the better-designed and more ubiquitous they become, the more I dislike them. I blame the customers. Awful people. Awful. Stop showing me your iPhone. Stop stroking your Macbook. Stop telling me to get one.

Seriously, stop it. I don't care if Mac stuff is better. I don't care if Mac stuff is cool. I don't care if every Mac product comes equipped with a magic button on the side that causes it to piddle gold coins and resurrect the dead and make holographic unicorns dance inside your head. I'm not buying one, so shut up and go home. Go back to your house. I know, you've got an iHouse. The walls are brushed aluminium. There's a glowing Apple logo on the roof. And you love it there. You absolute MONSTER.

Of course, it's safe to assume Mac products are indeed as brilliant as their owners make out. Why else would they spend so much time trying to convert non-believers? They're not getting paid. They simply want to spread their happiness, like religious crusaders.

Consequently, nothing pleases them more than watching a PC owner struggle with a slab of non-Mac machinery. It validates their spiritual choice. Recently I sat in a room trying to write something on a Sony Vaio PC laptop which seemed to be running a special slow-motion edition of Windows Vista specifically designed to infuriate human beings as much as possible. Trying to get it to do anything was like issuing instructions to a depressed employee over a sluggish satellite feed. When I clicked on an application it spent a small eternity contemplating the philosophical implications of opening it, begrudgingly complying with my request several months later. It drove me up the wall. I called it a bastard and worse. At one point I punched a table.

This drew the attention of two nearby Mac owners. They hovered over and stood beside me, like placid monks.

"Ah: the delights of Vista," said one.

"It really is time you got a Mac," said the other.

"They're just better," sang monk number one.

"You won't regret it," whispered the second.

I scowled and returned to my infernal machine, like a dishevelled park-bench boozer shrugging away two pious AA recruiters by pulling a grubby, dented hip flask from his pocket and pointedly taking an extra deep swig. Leave me alone, I thought. I don't care if you're right. I just want you to die.'

That final sentence probably hits the nub of the issue more effectively than any other. We can get ourselves so entrenched in our own social tribe that we just cannot countenance the perspectives of a different one. 'Why should we when our tribe is right and even if it is wrong, there are enough of us to make it seem all right?', the thinking seems to go. Rationality will always lose to emotion.

In the arena of climate change, there is a danger that the questions of rationality will lose out to the questions of emotion. Questions about scientific fact become discussed as issues of social power. Normally, that could just be seen as part of the rich tapestry of life: some win, some lose. But climate change is different. The climate trajectory we are currently on cannot produce winners. We will all lose. This is our evolutionary choice. We face up to our tribalism or face up to the mortality of our species. We are still in charge of our own destiny but we must choose to be winners.

We don't like to contemplate our own mortality and climate change probably forces us in the developed world to do so more than any other issue.

There is a joke about death that I rather like which goes: 'When I die I want to go in my sleep like my grandpa did, not screaming in panic, like the passengers in the car'. We don't know when our personal end will come but most of us hope it will at least not be too traumatic.

Humans are a surprisingly difficult bunch to kill in big numbers. The fastest rates seen so far are the 2004 tsunami which killed 250,000 in a few hours, which roughly equates to 2 million per day. The Rwandan massacres of 15 years ago saw 1 million people killed in 100 days, so an additional 10,000 deaths a day in a population that would ordinarily expect to see an average of 350. World War II killed an estimated 55 million over 8 years, which equates to an average of about 22,000 additional deaths per day at a time when the global daily death toll would ordinarily be somewhere in the region of around 45,000 per day. Today, the global daily death toll is about 125,000 per day and maybe 1,000 of those can be directly related to climate change. Some commentators have suggested that in some scenarios, climate change will cause billions to die over a period of just a few years. For this to happen is not impossible and will certainly be suffering on an unprecedented scale: it would be like suffering the 2004 tsunami all day, every day, over a period of years. If we allow things to go that far, we will not be the first civilisation to push our environment to collapse and cause our own decline but we will almost certainly be the first one to choose to do it knowingly.

I haven't devoted much time in the course of this book to looking at what might happen if we carry on with business as usual. Other writers have covered it extensively and Mark Lynas' *Six Degrees: Our Future on a Hotter Planet* is a detailed look at the kind of physical effects that the world may experience. The novel *A Song of Stone* by Iain Banks was not written about climate change, but gives a bleak assessment of how people can act when social order breaks down. The novel *The Rapture* by Liz Jensen is part of a growing canon of works that address climate change and depicts a heroine who finds herself facing a future in which there is 'no safe place for a child to play. Nothing but hard burnt rock and blasted earth, a struggle for water, for food, for hope. A place where everyday will be marked by the rude, clobbering battle for survival and the permanent endurance of regret, among the ruins of all we have created and invented, the busted remains of the marvels and commonplaces we have dreamed and built, strived for and held dear: food, shelter, myth, beauty, art, knowledge, material comfort, stories, gods, music, ideas, ideals, shelter.....A world I want no part of. A world not ours.'*

To switch the entire world away from its use of fossil fuel and towards renewable energy will certainly be a Herculean task, but it can be achieved. It simply takes a collective belief that we can do it. Your adoption of the Labours may be copied by your friends and neighbours. The understanding that we need to choose a different fuel will

* Reproduced with permission of Bloomsbury Publishing.

spread. Other books will say a similar thing in a different way. Television programmes will be made with the same message. Politicians will recognise that they have been empowered to take on the vested interest of those who make money from fossil fuels and force through a switch by a mix of political, legal and economic means. The collective good will win out over the greed of the individual. The world will get into its flow.

The idea of 'flow' or 'collective unconsciousness' is not a new one. It can be traced back to the ancient Chinese and their philosophy which saw the universe in a constant state of flux. Humans have the choice of understanding this flux and living in harmony with it by following 'the way', although they can also battle against it.

The ancient Greeks had a slightly different way of looking at things. Aristotle classified the essential elements as fire, earth, air and water, along with a fifth substance he referred to as 'aether' and regarded as a divine substance. The nature of aether has been speculated upon by many scientists and artists over the centuries. Early in the 20th century, Einstein unveiled his theory of relativity and a new view of the universe in which cause need not precede effect. Einstein, the master of matter, was a contemporary of Carl Jung, who was a master of the newly emerging study of the mind. The two met over a series of dinners in Zurich[104] and their discussions crossed between the disciplines of mind and matter, helping Jung develop his theory of a collective unconsciousness.

Jung's distinction between a conscious temporary world and an unconscious eternal one was supplemented by his concepts of 'synchronicity' or 'meaningful coincidence', which he saw as evidence of the unconscious world emerging into the conscious one. The idea of a collective unconscious is a tantalising one and there is now some intriguing evidence emerging to back up the theory.

In 1980, Dr. Roger Nelson, an experimental psychologist with a background in physics, statistical methods and engineering, joined a group at Princeton University in the US that had been looking at the interaction between people and electronic equipment. Over the years, he and his colleagues have been running what has become known as the Global Consciousness Project. This involves collecting data continuously from a global network of random number generating machines located in 65 host sites around the world, which feed their data back to a central database. The expectation, obviously, is that this database should simply be an enormous collection of random data. But that is not what appears to be happening. Using very strict criteria, Dr. Nelson began to examine patterns in the data as they correlated to major world events. Events studied included such things as the 9/11 terrorist attacks, the death of Princess Diana and even New Year's Eve celebrations. What Dr. Nelson discovered was that in the majority of these instances, the sampled data from the EGG network shows significant deviation from the expected patterns. Put another way, the results seem to indicate that for major global events which impinge upon the consciousness of a vast number of the world's people, there seems to be some influence on the EGG network: the random number generators begin to show meaningful departures from expectations. There may indeed be a measurable global unconsciousness.[105]

Some philosophers have concluded that there is no meaning to life other than to live it, but here we have an amazing prospect. What if the reward for overcoming our greatest challenge is that we take the next step in human evolutionary history? What if humankind who, in a blink of an evolutionary eye, have gone from learning to stand upright to being able to fly free of the bonds of the planet, really can go further? What if each of us is connected to everyone else through consciousness? What could we achieve then? All it will take for us to find out is for enough of us to make the positive choice.

China has now overtaken the United States as the world's largest emitter of greenhouse gases. China is not proud of that record and is fast emerging as the biggest driver for the change to renewable energy and is currently planting more trees each year on average than are being destroyed in Brazil. The United States, despite the election of a new President, is still finding it difficult to come to terms with the challenge. The United States has had a powerful impact on the world thus far, and how the US and China behave in the coming years will have a powerful impact on our future. The Copenhagen Accord managed only to specify what the world was going to try to achieve, not how it would achieve it. So to the leaders of the world, I would extend an invitation to make this declaration in 2010. I have named it the Arrhenius Rice Declaration.

'We choose to reinvent accountancy so that the Earth-cost of consumption is included in the price of production. In doing so, we will quickly lay aside the use of fossil fuel in favour of renewable energy. We choose to do this so the concentration of CO_2 can return to below 350 ppm, a level we know allows for a climate safe for humans. We choose to reinvent accountancy in this decade so that our civilisation may prosper in decades to come. We choose to reinvent accountancy not for our children's sake but for our sake, not because it is easy, but because it is hard, because that goal will serve to organise and measure the best of our energies and skills, because that challenge is one that we are willing to accept, one we are unwilling to postpone, and one which we intend to win.'

Can we do it? Having borrowed the words of one American President, I will do it again: 'Yes we can'. For I believe in the words of the poet Marianne Williamson:

'Our deepest fear is not that we are inadequate.

Our deepest fear is that we are powerful beyond measure.'

I hope that this book has given you the belief and tools you need. What our world needs now is more people to become a Humankind Superhero.

References

Where a reference is given as a web address, it was correct at the time of this book's completion (December 2009), but may have subsequently been moved by the website's owner.

Foreword
1. This report is available free of charge at www.tinyurl.com/5qfkaw
2. 'Plastic bags being used less, say retailers', *Guardian* (1 May 2009). http://www.defra.gov.uk/ENVIRONMENT/localenv/litter/bags/

Part One
3. http://www.makingthemodernworld.org.uk/stories/defiant_modernism/01.ST.02/?scene=6&tv=true
4. http://www.le.ac.uk/ge/genie/vgec/sc/dna.html
5. Adam Smith, *The Wealth of Nations* http://www.adamsmith.org/smith/won-intro.htm
6. Wrigley, E. Anthony, and Roger S. Schofield, *The Population History of England, 1541–1871: A Reconstruction* (Harvard University Press, Cambridge, 1981). Office for National Statistics Press Release 30/10/08 'Life expectancy at birth...'
7. Quoted in 'Eager sellers, stony buyers' *Harvard Business Review* (June 2006).
8. Davies, Glyn. *A History of money from ancient times to the present day*, (University of Wales Press, Cardiff, 2002)
9. Peter Bernstein, *Against the Gods* (John Wiley & Sons, 1996)

Part Two
10. www.just-auto.com (July 24th 2009)
11. ONS reported in *Daily Telegraph* (24 July 2009).
12. http://www.esrl.noaa.gov/gmd/aggi/
13. IPCC Climate Change 2007: 'Synthesis Report Summary for Policymakers'
14. http://news.bbc.co.uk/go/pr/fr/-/2/hi/americas/3582794.stm
15. John Lanchester's book about the financial crisis, *Whoops*, will be published by Penguin Press, 'once he's finished writing it' LRB (28 May 2009).
16. 'Killing Speed and Saving Lives', UK Dept. of Transportation, London, England. See also Limpert, Rudolph. *Motor Vehicle Accident Reconstruction and Cause Analysis*, (The Michie Company, Charlottesville VA.1994)
17. George Monbiot, *'Heat'* Penguin 2007
18. http://news.bbc.co.uk/1/hi/uk/7628137.stm
19. http://blogs.news.sky.com/cityblog/Post:a7314dcf-64a9-4e0c-adfc-66ed829cdbec
20. 'Psychology and Global Climate Change: Addressing a Multi-faceted Phenomenon and Set of Challenges', a report By The American Psychological Association's Task Force On The Interface Between Psychology And Global Climate Change
21. Chris Stringer & Robin McKie *African Exodus: The Origins of Modern Humanity* Jonathan Cape 1996
22. http://www.ausbcomp.com/~bbott/cars/carhist.htm
23. 'Affective Motives For Car Use', Linda Steg, Centre for Environmental and Traffic Psychology, University of Groningen, Gerard Tertoolen, Transport Research Centre AVV
24. National Travel Survey, 2008, Department for Transport
25. Author's calculation from DfT data, 2006
26. Transport Statistics 2008, DfT
27. David J.C. MacKay, *Sustainable Energy – without the hot air* (UIT, Cambridge, 2008) Available free online from www.withouthotair.com it gives an excellent comparison of the energy cost of various activities
28. National Travel Survey 2008
29. A full version of the interview is available at www.tinyurl.com.2ykfgw
30. 'Johnson ups carbon footprint by courtesy of flight to New York, *Guardian* (15 September 2009)
31. http://www.chooseclimate.org/flying/mf.html provides a great tool to calculate any permutation you might like on flying

32. This price was available at www.cheapflights.co.uk for travel at the end of October 09
33. http://www.timesonline.co.uk/tol/travel/news/article623062.ece
34. http://www.telegraph.co.uk/news/uknews/1545807/You-cant-change-world-by-wearing-sandals.html
35. McDonough and Braungart ,*Cradle to Cradle* (North Point Press, 2002)
36. DEFRA
37. Environmental Benefits of Recycling, WRAP, 2006
38. http://www.berr.gov.uk/files/file15400.pdf
39. http://www.carbontrust.co.uk/climatechange/policy/green-tariffs.htm
40. http://www.guardian.co.uk/media/2008/jan/30/asa.advertising1?gusrc=rss&feed=media
41. National Statistics, Family Spending 2008
42. http://www.ukgbc.org/site/news/showNewsDetails?id=104
43. http://www.energysavingtrust.org.uk/business/Business/Building-Professionals/Existing-housing
44. http://www.ukgbc.org/site/news/showNewsDetails?id=104
45. http://www.defra.gov.uk/environment/business/reporting/conversion-factors.htm
46. http://www.seeit.co.uk/haringey/Map.cfm
47. http://www.energysavingtrust.org.uk/Home-improvements-and-products
48. http://www.ecomerchant.co.uk/index.php?main_page=n index&cPath=8&zenid=07135037c13458a95b3f784c0c6f1e8b
49. http://www.passive-house.co.uk/passive_house.htm
50. http://www.byebyestandby.com/homeindex.php
51. http://www.energysavingtrust.org.uk/Compare-and-buy-products
52. http://www.energysavingtrust.org.uk/Energy-saving-assumptions
53. http://www.homeenergysaving.co.uk/electricity-monitors.html?source=aw_electricitymonitor_lg_grp&gclid=Cli-wPnCj5wCFUYA4wod70JjYw
54. http://decc.gov.uk/en/content/cms/consultations/elec_financial/elec_financial.aspx
55. http://www.carbonfreegroup.com/photo-voltaic-thermal-panels.html
56. http://www.newformenergy.com/solar-thermal.html
57. http://www.ecocentre.org.uk/solar-hot-water.html

58. http://www.paynesheatcentre.co.uk/solarth.htm
59. http://www.paynesheatcentre.co.uk/a2w.htm
60. http://www.dimplex.co.uk/products/renewable_solutions/heat_pump_air_source/outdoor/la_mr/index.htm
61. http://www.carbonfreegroup.com/geo-thermal-storage.html
62. http://news.bbc.co.uk/1/hi/uk/6731659.stm
63. http://www.heraldscotland.com/food-for-thought-why-home-cooking-has-soared-as-the-recession-bites-hard-1.902279
64. WRAP; The Food We Waste
65. http://www.imaner.net/panel/statistics.htm
66. See 'Global Warning: Climate Change And Farm Animal Welfare', a report by Compassion in World Farming, 2008, for a good summary of the issues and where most of the data quoted in this chapter is from.
67. http://www.ens-newswire.com/ens/sep2009/2009-09-08-03.asp
68. http://www.telegraph.co.uk/finance/newsbysector/retailandconsumer/5183416/Online-shopping-soars-in-popularity-as-Brits-avoid-supermarket-hassle.html
69. *Shopping: The Pleasure/Pain Principle*, PFSK Feb 09
70. David J.C. MacKay, *Sustainable Energy – without the hot air* (UIT Cambridge, 2008)
71. PriceWaterhouseCoopers LLP; Sustainability: Are Consumers Buying It?
72. http://www.climatechangecorp.com Carbon labels – a green mark too far?
73. Professor Richard Wiseman, *59 Seconds: Think a Little Change a Lot* (Macmillan, 2009)
74. Omar Vidal and Jorge E. Illueca,*Transfer of Environmentally Sound Technologies for the Sustainable Management of Mangrove Forests: An Overview* (Worldwide Fund for Nature, Mexico, 2008)
75. http://www.ucsusa.org/global_warming/solutions/forest_solutions/international-timber-trade.html
76. 'Loch Katrine, Scotland' *Guardian* (14 December 2009)
77. The Prince of Wales' Rainforests Project, An Emergency Package for Tropical Rainforests, March 2009
78. United Nations Forum on Forests, Eighth session, New York, 20 April-1 May 2009.
79. Nasa Earth Observatory, Tropical Deforestation Report.
80. Forestry Commission, Climate Change and British Woodland
81. http://tinyurl.com/cochainvest
82. http://www.aip.org/history/climate/xVostokCO2.htm

83. www.corpwatch.org
84. Fortune 500, 2008
85. Dow Chemicals 2007 Global Reporting Initiative Report
86. Dow Chemical 2015 Sustainability Report Q209
87. http://www.asa.org.uk/asa/adjudications/Public/TF_ADJ_44828.htm
88. Data available at www.catlinarcticsurvey.com
89. Available at http://climatesafety.org/downloads/
90. http://www.esrl.noaa.gov/gmd/ccgg/trends/
91. A Survey of the UK Tax System, Institute for Fiscal Studies Briefing Note No. 9
92. 'Tax on Coca-Cola 'could make Americans thinner'' *Observer* (4 October 2009)
93. IFS study as above
94. *Plan B 4.0 Mobilising to save civilisations*, Lester R Brown, Earth Policy Institute 2009
95. Ibid
96. http://energy.gov/energysources/coal.htm
97. 'Growth in opencast coal mining dismays climate campaigners' *Guardian* (15 August 2009)
98. Al Gore, *Our Choice* (Bloomsbury, 2009)
99. 'Scientists explore how the humble leaf could power the planet' *Guardian* (11 August 2009)
100. David J.C. MacKay, *Sustainable Energy – without the hot air* (UIT, Cambridge, 2008). Available free online from www.withouthotair.com
101. 'Some 600,000 join millionaire ranks in 2004', Eileen Alt Powell, (Associated Press, 9 June 2005). Wikipedia list of billionaires 2009
102. Global Climate Change Policy Tracker: An Investor's Assessment, October 2009 Deutsche Bank
103. The lower figure is from Plan B, the higher from the *Guardian* (23 November 2009)

Part Three
104. Roger Hamilton, *Your Life, Your Legacy* (Achievers International, 2006)
105. See http://noosphere.princeton.edu/ for details of the data and hypothesis

Bibliography

In addition to those books specifically referenced, I have also found the following enlightening and stimulating.

Carbon Detox, George Marshall, Octopus Publishing Group Ltd 2007
A Blueprint for a Safer Planet, Nicholas Stern, The Bodley Head 2009
Payback, Margaret Atwood, Bloomsbury 2008
The Greek Myths, Volume II, Robert Graves, The Folio Society 2002
Facing the World with Soul, Robert Sardello,
Against the Gods, The Remarkable Story of Risk, Peter L Bernstein, John Wiley & Sons, 1998
Greed: Why we can't help ourselves, Richard Girling, Doubleday 2009

Appendix 1

Addresses for the Head Offices of Asda, Sainsbury's, Tesco and Dow Chemical:

Asda ASDA House,
Southbank,
Great Wilson Street
Leeds LS11 5AD

J Sainsbury plc
33 Holborn
London EC1 N2HT

Tesco PLC
New Tesco House
Delamare Road
Cheshunt
Hertfordshire
EN8 9SL

The Dow Chemical Company,
2030 Dow Center,
Midland, MI 48674
USA

Appendix 2

The 100 most marginal constituencies: 2005 election.

Aberdeen South
Angus
Battersea
Bethnal Green and Bow
Birmingham Edgbaston
Bolton West
Broxtowe
Burton
Calder Valley
Cardiff North
Carmarthen West and Pembrokeshire South
Carshalton and Wallington
Ceredigion
Chatham and Aylesford
Cheltenham
City of Chester
Clwyd West
Colne Valley
Corby
Croydon Central
Crawley
Dartford
Dorset North
Dorset South
Dorset West
Dumfriesshire, Clydesdale and Tweeddale
Dundee East
Eastbourne
Eastleigh

Edinburgh North and Leith
Edinburgh South
Enfield North
Enfield Southgate
Falmouth and Camborne
Finchley and Golders Green
Forest of Dean
Gillingham
Gravesham
Guildford
Harlow
Harrow West
Harwich
Hastings and Rye
Hemel Hempstead
Hereford
High Peak
Hornchurch
Hornsey and Wood Green
Hove
Ilford North
Islington South and Finsbury
Leeds North West
Loughborough
Ludlow
Manchester Withington
Medway
Milton Keynes North East
Na h-Eileanan an Iar
Nuneaton
Ochil and South Perthshire
Oxford East
Pendle
Perth and North Perthshire
Portsmouth North
Preseli Pembrokeshire
Putney
Reading East
Ribble South
Rochdale
Romsey
Rugby and Kenilworth

Scarborough and Whitby
Selby
Shipley
Shrewsbury and Atcham
Sittingbourne and Sheppey
Solihull
Somerton and Frome
St Albans
Stafford
Staffordshire Moorlands
Stourbridge
Stroud
Swindon South
Taunton
Thanet South
The Wrekin
Torbay
Totnes
Vale of Glamorgan
Wansdyke
Warwick and Leamington
Watford
Wellingborough
Westmorland and Lonsdale
Weston-Super-Mare
Wimbledon
Wirral West
Worcestershire West
Ynys Mon

Acknowledgements

Books are in some ways like symphonies: they may have a single author, but they need many players to make them come alive. I am indebted to the help I have received from many people, directly and indirectly, that has helped me to shape this work. Special thanks need to go to my wife, Karin, for her unstinting support and extensive knowledge of language, and my mother in-law, Arna, for her insights to the human condition. Thanks also to the people who have helped shape the idea and review the text: Ben Evans, David Howe, Charlie Roberts, Robert Copping, Jakob von Baeyer, Adrian Drewe, Liz Jensen, Jae Mather, Mike Scase, David Corr, Richard Guest, Muriel Barrie, Ruth Betts and Vanessa Harvard Williams. I am also deeply indebted to Mary Archer, not just for her comments on the text but also, by agreeing to share a pot of tea, changing the way I looked at the world. Thanks also to Good Energy for providing the promotion.